18

The Institute of Mathematics
and its Applications
Conference Series

The Institute of Mathematics
and its Applications
Conference Series

Previous volumes in this series were published by
Academic Press to whom all enquiries should be addressed.
Forthcoming volumes will be published by
Oxford University Press throughout the world.

NEW SERIES

1. *Supercomputers and parallel computation* Edited by D. J. Paddon

Supercomputers and parallel computation

Based on the proceedings of a workshop on
Progress in the Use of Vector and Array Processors
organized by The Institute of Mathematics and its Applications
and held in Bristol, 2–3 September 1982

Edited by

D. J. PADDON
University of Bristol

CLARENDON PRESS · OXFORD · 1984

Oxford University Press, Walton Street, Oxford OX2 6DP

London New York Toronto
Delhi Bombay Calcutta Madras Karachi
Kuala Lumpur Singapore Hong Kong Tokyo
Nairobi Dar es Salaam Cape Town
Melbourne Auckland

and associates in
Beirut Berlin Ibadan Mexico City Nicosia

Oxford is a trade mark of Oxford University Press

Published in the United States
by Oxford University Press, New York

British Library Cataloguing in Publication Data
Supercomputers and parallel computation. – (The Institute of Mathematics and its Applications Conference series. New series; 1)
1. Parallel processing (Electronic computers)
I. Paddon, D. J. II. Institute of Mathematics and its Applications III. Series
001.64 QA76.6
ISBN 0-19-853601-1

Printed in Great Britain by St Edmundsbury Press, Bury St Edmunds, Suffolk

PREFACE

The late 1970's marked the arrival of a number of commercially available vector and array processing computers. A world-wide interest in research relating to the design and the use of these computers has rapidly developed. In 1981 the Japanese Government announced the 'fifth generation computer systems' research programme, whose main aim was to develop computer systems for the next decade. The implications of the Japanese research programme caused both excitement and alarm throughout the rest of the world and in response many nations have announced their own 'fifth generation' programmes. The main architectural feature of this next generation of super-computers will be their inherent parallelism and accordingly, since 1981, even greater impetus has been given to the parallel computing research community.

This volume is based on the proceedings of a conference on parallel computing held at the University of Bristol in September, 1982. The proceedings are organised in such a way that related areas of research appear together. The order of topics is broadly: VLSI parallel architectures, theory of parallel computing, vector and array processor computing.

Most of the experience gained in array processor computing in the United Kingdom has been associated with the ICL distributed array processor (DAP). These proceedings reflect this experience. However, most of the research relating to the ICL DAP is of fundamental importance to the design and use of all types of array processors.

I would like to express my appreciation to the co-organiser of the conference, Dr. J.D. Pryce of the University of Bristol, and to the conference sponsor Floating Point Systems (UK) Ltd. My appreciation is also expressed to the publishers and to the staff of the Institute of Mathematics and its Applications whose care and enthusiastic help were necessary for this publication.

D.J. Paddon
University of Bristol

ACKNOWLEDGEMENTS

The Institute thanks the authors of the papers, the editor, Dr. D. Paddon (University of Bristol) and also Miss Denise Wright and Mrs. Janet Parsons for typing the papers.

CONTENTS

1. VLSI architectures for problems in numerical computation by P.M. Dew — 1

2. Path Pascal simulation of multiprocessor lattice architectures for numerical computations by M. Berzins, T.F. Buckley and P.M. Dew — 25

3. A reconfigurable processor array for VLSI by C.R. Jesshope — 35

4. Examples of array processing in the next FORTRAN by A. Wilson — 41

5. Optimizing the FACR(ℓ) Poisson-solver on parallel computers by R.W. Hockney — 45

6. Program development software for array processors by J.M. Marsh — 67

7. On making the NAG run faster by D.H. McGlynn and L.E. Scales — 73

8. Tough problems in reactor design by W.D. Collier, C.W.J. McCallien and J.A. Enderby — 91

9. On the design and implementation of a package for solving a class of partial differential equations on the ICL distributed array processor by S.L. Askew and F. Walkden — 107

10. The three-dimensional solution of the equations of flow and heat transfer in glass-melting tank furnaces: adapting to the DAP by A.F. Harding and J.C. Carling — 115

11. Document abstracting on the distributed array processor by D.E. Oldfield — 135

12. Sparse matrix vector multiplication on the DAP by R.H. Barlow, D.J. Evans and J. Shanehchi — 147

13. Unstructured sparse matrix vector multiplication on the DAP by M. Morjaria and G.J. Makinson — 157

14. Band matrices on the DAP by L.M. Delves, A.S. Samba and J.A. Hendry — 167

15. GEM calculations on the DAP by J.A. Hendry and L.M. Delves — 185

16. The implementation of the FFT on the DAP by S.T. Davies — 195

17. QU factorisation and singular value decomposition on the DAP by J.J. Modi and G.S.J. Bowgen — 209

18. Implementation of a parallel (SIMD) modified Newton algorithm on the ICL DAP by K.D. Patel 229

Reference Index 251

Index 255

VLSI ARCHITECTURES FOR PROBLEMS IN NUMERICAL COMPUTATION

P.M. Dew

(Department of Computer Studies, The University, Leeds)

ABSTRACT

Recent advances in VLSI technology have led to a rapid growth in the research into parallel computer architectures. The purpose of this paper is to consider the application of these architectures to problems arising in numerical computation. A general introduction to parallel architectures and VLSI design is followed by an example illustrating the use of a special-purpose architecture for a real-time control problem. Some of the more interesting architectures proposed for matrix computations are considered including the computational wavefront machine, the finite element machine, and systolic array processors. Finally the design of a systolic array for the Jacobi Iterative method is considered.

1. INTRODUCTION

The need to perform ever more complex scientific computations is clearly illustrated by the success of the new generation of vector processors like the CRAY1. The speed-up achieved by these super-computers has, in the main, been brought about by architectural improvements (e.g. pipelining floating point operations) and the development of compilers to map programs designed for serial machines onto a vector architecture. Although new algorithms are being developed to exploit the features of machines like the CRAY1, in general, the introduction of vector processors has had very little effect, apart from minor tuning, on the underlying numerical algorithms. This is not the case for the new generation of array processors like the ICL DAP computer, as clearly illustrated by several contributions to this Workshop.

The general theme of this paper, is that to obtain still further speed-ups the problem, the algorithm, and the architecture must be more closely related. This is illustrated by the work of S. Pawley and his group (Pawley and Thomas (1982)) who are studying how molecules rearrange themselves with changing state. They estimate that by using the ICL DAP computer (and hence mapping more closely the architecture to the problem) the cost of the experiment in terms of computer hardware and time was just 4% of the CRAY1. A major attraction of the emerging VLSI technology is that it will give us the potential to design, in a cost effective manner, devices tailored to particular problem classes (e.g. solution of elliptic partial differential equations discussed later in this paper).

There has been a rapid expansion of the research into highly parallel and specialised computer architectures to exploit the potential of VLSI systems. An excellent 'state of the art' survey can be found in the January 1982 issue of IEEE Computer magazine. Packing densities of tens of thousands of transistors per chip (LSI) are possible today and into the future (10-20 years) factors of 10 to 100 times this amount (VLSI) are predicted. Of equal importance to the increase in packing density is the rapid development of sophisticated CAD tools which will drastically cut down the design time and make the technology available to a much larger user community.

An important development in the design of parallel computer architectures is the development of sophisticated microprocessors that can be "plugged" together to form a parallel processor (e.g. Systolic Processor Chip, Kung (1982), and the Transputer being developed at Inmos). As the packing density increases still further and the reliability improves (resulting in a larger area) so it will be possible to layout several processors on a chip. The trick then will be to arrange the processors and memories on the chip to suit the characteristics of the computation.

1.1 VLSI Parallel Architectures

The requirements of parallel architectures for VLSI have been discussed by many authors and we refer the reader to Kung (1982) and Seitz (1982). Briefly they are:

(1) Simple and Regular Design: the design should contain a few modules which are replicated many times. The complexity (or grain) of the modules depends on the application.

(2) Parallelism: the design must have a very high degree of parallelism through both pipelining and multiprocessing.

(3) Communication and Switching: the major differences between VLSI design and the earlier digital technologies is that the communication paths, relative to switching, will dominate both the area and the time delay (Seitz (1982)). This is because the speed of the devices increases as the feature size decreases while the propagation time along a wire does not. Successful algorithms for VLSI design will be the ones where the communication is only between neighbouring processors.

1.2 Outline of the Paper

A survey on the developments in the design of highly parallel computer architectures for numerical computation is given in section 2. A multiprocessor lattice architecture and systolic array architectures are considered in detail and an application of a systolic array architecture to a real-time control problem is given. Section 3 is mainly concerned with the design of special-purpose architectures for the solution of linear equations arising from the discretisation of partial differential equations.

2. DEVELOPMENTS IN PARALLEL COMPUTER ARCHITECTURES

For an up-to-date and interesting taxonomy of computer architectures the reader is referred to Seitz (1982) and for a general survey to Haynes, Lau, Siewiorek and Mizell (1982). The three broad classes that are of interest for numerical computations are:

(1) general-purpose parallel architectures,

(2) attached special-purpose processors, and

(3) signal processing type applications.

2.1 General-Purpose Parallel Architectures

The emphasis here is on designing large mainframe machines that have varying degrees of parallelism in a multi-user environment. Major research areas are the design of array processors, data flow machines, multiple processors like CM* and dynamically reconfigurable (i.e. under software control) array processors (see the bibliography in Haynes, Lau, Siewiorek and Mizell (1982) for published work on these developments and Love (1982)). The design of a suitable computer language to exploit the parallelism of these machines is of prime importance. In this paper, we wish to move away from large general-purpose processors and consider the progress that has been made in the design of more specialised processors.

2.2 Attached Special-Purpose Processors

It is an attractive idea to attach a special-purpose parallel processor to the system bus of a microcomputer like the ICL PERQ to speed-up the more computationally bound tasks. The type of arrangement normally envisaged is shown in Fig. 1. The special-purpose processor, for example, could be designed to exploit the features of a particular class of problems (e.g. the finite element machine (Podsiadlo and Jordan (1981)) or alternatively it could be designed for a general computational task, like the solution of linear equations, as envisaged by Kung (1979).

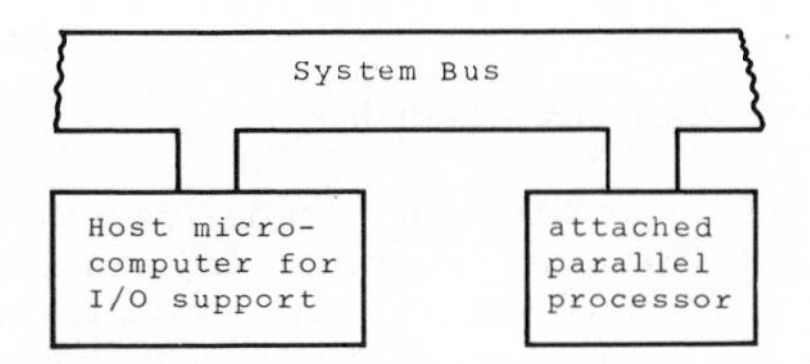

Fig. 1 Microcomputer with attached special-purpose processor

The general idea of a specialist work-station for the solution of elliptic partial differential equations, using a dual processing system, has been carefully considered in a paper discussing the prospects of a "high technology elliptic solver" (Eisenstat and Schulz (1981)). However, the I/O communication between the host and the attached processor is a serious problem and can severely limit its predicted usefulness. An important question is therefore the identification of key sub-algorithms that an attached processor can perform such that the gain is computational speed is not cancelled out by the movement of data. Several ways of dealing with the communication issue are discussed in Eisenstat and Shultz (1981) for an elliptic p.d.e. solver. Briefly these are

1. hide the I/O communication by ensuring that useful computation is overlapped with I/O communication,

2. trade off storage for computation time, e.g. by using linear algebra techniques to reduce the amount of storage required at the expense of computation time, and

3. use analytic techniques such as devising algorithms that use less data points (e.g. high order adaptive algorithms, see Ridgway-Scott (1981)).

In graphics applications, the problem of displaying the results can be overcome by arranging that part of the attached processor's memory also forms the pixel cells for a graphics display device. (Such an arrangement is envisaged for the recently announced ICL PERQ with DAP computer system.)

It is well known that oil and gas reservoir simulations are computationally very expensive and it would appear to be advantageous to use an attached processor to solve the linear equations involved in these calculations. However, because a significant amount of the computation time is spent in accessing large data bases to assemble the linear equations, it is unlikely that simply providing a processor for solving the equations will lead to significant speed-ups. Architectural improvements are also going to be needed to speed-up the assembly of the equations.

There are two general architectural designs that we shall consider in this paper for the design of the attached parallel processor. The first is a multiprocessor lattice architecture based on the idea of several processing elements operating under a centralised control and the second is systolic array architectures which make extensive use of pipelining but at a higher functional level than is used in a vector machine.

2.2.1 Multiprocessor Lattice Architecture

A multiprocessor lattice architecture as defined in Dew, Buckley and Berzins (1983), is a NxN array of processing elements which execute concurrently under a centralised control and communicate along local communication paths connecting neighbouring processing elements. Each processing element has sufficient *local memory* to store both the results and temporary values that may be needed during the computation. In addition the processing elements may communicate via some form of global bus but this is not an essential feature of the architecture. The ICL DAP computer is an example of a multiprocessor lattice architecture. Another example is the *Configurable Highly Parallel Computer (CHiP)* (Snyder (1982)) where *programmable switches* are provided between the processing elements. Being able to program the switches means that the lattice is reconfigurable dynamically and this opens up the interesting prospect of designing algorithms that change their communication pattern during the computation. Two general multiprocessor lattice architectures, namely the *computational wavefront machine* and the *finite element machine* are considered in section 3. When assessing the usefulness of this type of architecture, it is important to remember that each processing element is normally programmed via the host computer and this requires a significant amount of software support.

The need for a large local memory, together with the need to broadcast a program to each element, and the possible use of a global bus means that the architecture is less well suited than the systolic architectures to the requirements of a VLSI system. Nevertheless, the development of processing elements that can be "plugged" together with a minimum of support circuits, made possible by using VLSI systems, will drastically cut the cost of building the hardware. The general question of placing more than one processor on a chip, and the likely shortage of I/O pins, is considered in DeRuyck, Snyder and Unruh (1982).

2.2.2 Systolic Array Processors

One of the more interesting and novel ideas to emerge from VLSI research into parallel architectures, is that of systolic array processors. They were developed by H.T. Kung and his co-workers at Carnegie-Mellon University, and consist of a set of interconnected cells (processing cells) each capable of performing a "hardware" simple arithmetic operation (e.g. for simple matrix computations, each cell performs the inner product step c := c + axb - see Fig. 2a). The cells are connected to form a 1 or 2 dimensional pipeline; once the pipeline is full the results arrive at constant time.

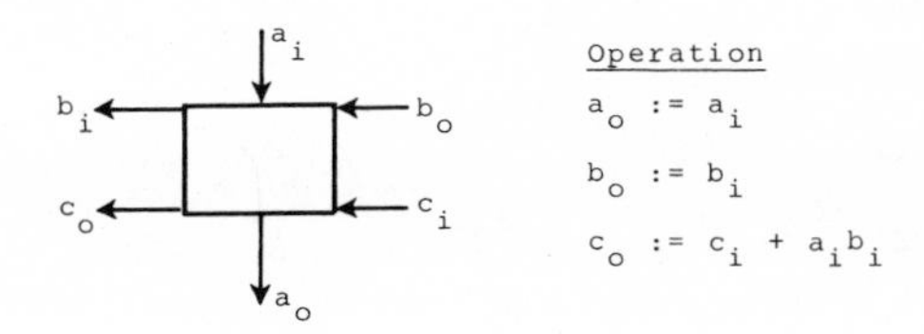

Fig. 2a Inner-product step processing cell

The important distinctions between a systolic array processor and the multiprocessor lattice are that in a systolic array

(i) the communication is only to neighbouring processing cells (i.e. no global bus).

(ii) communication with the outside world occurs only at the boundary cells (i.e. the processing cells at the edges of the lattice), and

(iii) the processing cells are "hardwired" (e.g. using firmware) and not programmed from the host.

Systolic array architectures satisfy the basic requirements for a VLSI system, although the need to broadcast control information could prove to be a bottleneck.

An excellent introduction to the basic principles of systolic arrays for matrix computations is given in Mead and Conway (1980). In Fig. 2b, we illustrate the Kung-Leiserson systolic array for the matrix/vector multiplication

$$\underline{y} = A\underline{x}$$

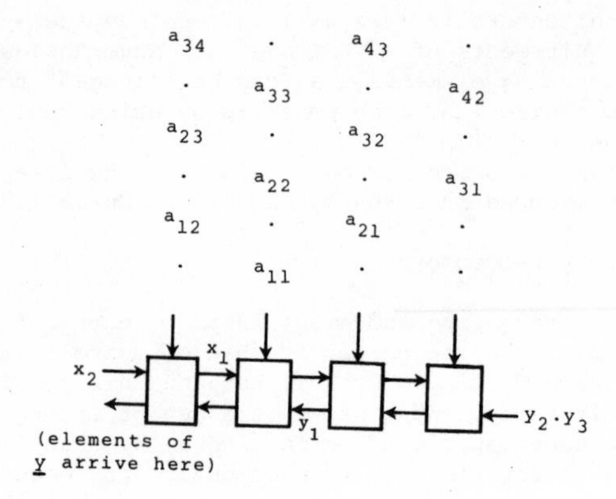

Fig. 2b Kung-Leiserson systolic array for banded matrix multiplication

where A is a banded matrix of the form

$$\begin{bmatrix} a_{11} & a_{12} & & \bigcirc \\ a_{21} & a_{22} & a_{23} & \\ a_{31} & a_{32} & a_{33} & a_{43} \\ \ddots & \ddots & \ddots & \ddots \\ \bigcirc & & & \end{bmatrix} \tag{2.1}$$

The elements of the vector $\underline{x}$, flow from left to right, the matrix coefficients flow into the top and the solution elements appear from the left. In this example half the processing cells are active at any one time, although it is possible to adjust the data flow so that they are all active (Kung (1982)). It is important to notice that the data is flowing regularly through the processor; this is an essential feature of any systolic array architecture and from which it derives its name. Also notice that the number of processing cells is dependent on the bandwidth of A and not on the size of A.

Systolic arrays are known for a wide variety of mathematical computations and an up-to-date account of these developments at Carnegie-Mellon University and elsewhere can be found in Kung (1982). Of special interest is the programmable systolic chip (PSC) that is currently being designed at CMU. The basic idea is that it will have sufficient memory to store a limited amount of data and will operate under a microprogram. The chip has six I/O busses to allow local neighbour communications.

The microprogram determines both the communication paths and the particular computation carried out by the cell.

The formal specification of algorithms for systolic arrays has recently been considered by Leiserson and Saxe (1981) and by Rogers (1982).

2.2.3 The CMU Systolic System

Simply connecting a systolic array processor to a system bus is unlikely to result in a significant speed-up because the data bandwidths for systolic arrays are very wide and so the communication cost would become prohibitive. Another interesting development at CMU is a system to support the systolic arrays. The basic functions blocks are shown in Fig. 3. An interface processor communicates with the host and has a relatively large memory to store the intermediate results and facilitate the chaining of the systolic array processors (SAP). The SAP handlers (SAPH) are microprogrammable processors that interface with the SAPs. This means that the support architecture can be very general and it is the selection of the special-purpose SAPs that tailors the system to a particular problem class. It would seem that the CMU Systolic System is very suitable architecture for supporting a hardware numerical library.

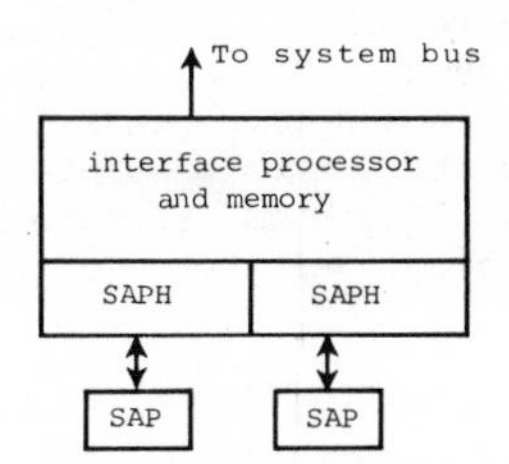

Fig. 3 The CMU Systolic System

The main architectural features of vector machines like the CRAY1 are incorporated in the CMU Systolic System. These are

(a) multifunction: the CRAY1 has a number of function units to perform basic operations like floating point addition and floating multiplication; the systolic system is also multifunction although the functions perform operations at a higher level (e.g. matrix factorisation).

(b) pipelining the function units of both systems are pipelined, and

(c) chaining the CRAY1 supports vector-chaining through a series of vector registers; this means that the output of one unit is fed directly into the input of another unit (see p. 120 of Ibbett (1982) for further details). Similarly the systolic system will support chaining of its systolic array processors.

The problem of programming such a system, and also how to handle matrices that become too large for a particular systolic array processor are briefly considered in Kung (1982).

2.3 Signal Processing Type Applications

This is a rich source of novel architectures because the data flow is regular, there is a need to perform computationally tasks in real time, and the applications are sufficiently general to justify building special-processors. Useful introductory references are Ahmed, Delosme and Morf (1982) and Denyer (1982). The systolic array architectures discussed above are ideal for signal processing applications, as illustrated in the following example.

Dr. DePennington, jointly with the Mechanical Engineering Laboratory at GEC and a Research Student John Finney, is investigating the feasibility of using digital control for the motion of a heavy mass on an almost frictionless surface (this work has important applications to machine tooling and robotics). Essentially they are using a self tuning controller to send electrical signals to an electrohydraulic cylinder drive. Self-tuning controllers have been very successful for systems with a slow response but have not been used for systems with a rapid response like the one described above, because of the large amount of "number-crunching" required.

The basic idea of a self tuning control algorithm (see Fig. 4a), is that once sufficient sampling data has been received, a set of model parameters are computed using a variant of the least squares method known as a square root filter Peterka (1975). (The square root filter was developed to improve the numerical characteristics of Kalman filters which are widely used in adaptive control algorithms.) Having obtained the model parameters it is then possible to determine a recursive formula, based on the sampling data, for the new control signal. For example, in the above problem, this formula is given by

$$y_i = B_1 y_{i-1} + (A_0 x_i + A_1 x_{i-1} + A_2 x_{i-2}) + H \qquad (2.2)$$

where y_i is the control signal and x_i is the output from the system (i.e. the position vector). The coefficients B_1, A_0, A_1 and A_2 depend on the model parameters and can be assumed constant for a particular configuration of the system, but this is generally not the case. John Finney has successfully implemented the self-tuning algorithm using a LSI11 microcomputer but was limited by the response of the computer. Further details can be found in Finney (1983).

The computation (2.2) is ideally suited for systolic array architecture as shown in Fig. 4b (see also Kung (1979)). The basic cell of the systolic array is an inner-product step processor which could be built using off-the-shelf multiplier/accumulator chips provided that we restrict the computations to fixed-point arithmetic (it significantly complicates the processor design to use floating-point computations).

The use of a systolic array to compute the basic control formula means that there is now no longer any need to keep the coefficients constant for the whole of the control period since these could be recomputed periodically, using a separate microprocessor. It remains an interesting challenge to design a systolic array to compute the model parameters using the square root filter equations. A microprocessor would then simply be needed to control the data flow through

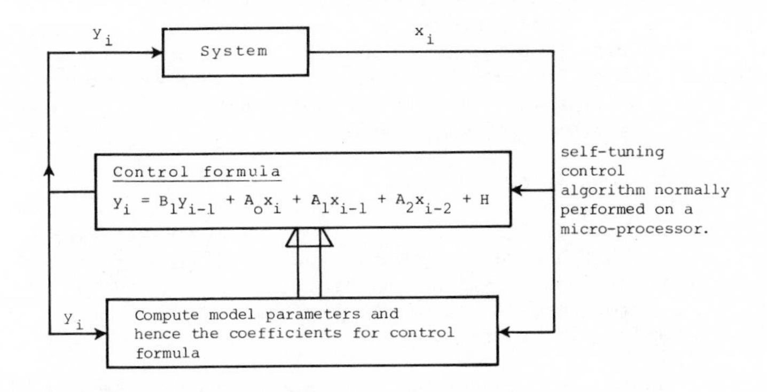

Fig. 4a Self Tuning Control

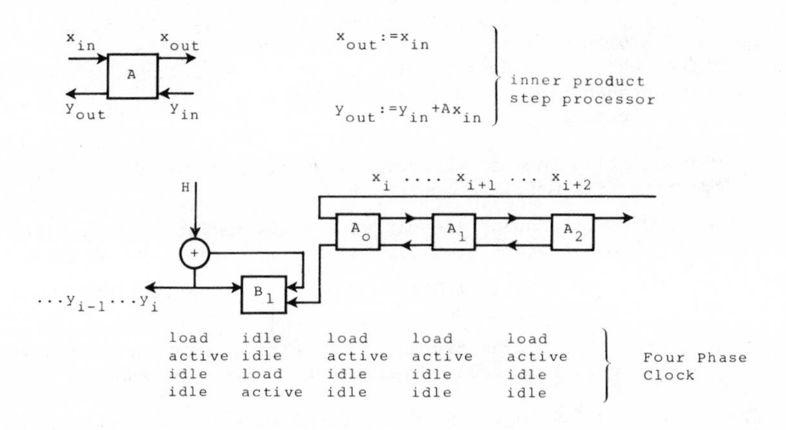

Fig. 4b Systolic Array for the Control Algorithm

the systolic architecture. The main problem here is likely to be that the calculation is not well-conditioned numerically so it may be necessary to use rather long word lengths.

A VLSI implementation of the systolic architecture will mean that complex self tuning controllers can be used even on systems requiring a rapid response.

3. SPECIAL-PURPOSE PROCESSORS FOR MATRIX COMPUTATIONS

A common feature of problems in scientific computing is the need to perform matrix arithmetic and it is not surprising that there is a large and growing body of literature concerned with the design of special-purpose processors and VLSI structures in particular, for this class of problems (see Mead and Conway (1980), Ahmed, Delosme and Morf (1982), and Kung (1982)).

In this section, we shall consider multiprocessor lattice and systolic array architectures for the solution of linear equations using both direct and indirect methods. The lattice of processing elements (or cells) are of a fixed size so when the number of equations becomes too large for the lattice additional time is required to partition the equations (Bokhari (1981)).

3.1 Wave Front Array Processor

This machine developed by S.Y. Kung and his co-workers (Kung, Gal-Ezer and Arun (1982)) has been especially designed for matrix computations. The important feature of this machine is the *computation wavefront notation* which removes the need for global synchronisation and also proves useful for programming the machine and describing the algorithms.

The machine architecture is essentially the multiprocessor lattice with additional memory modules on the north and west edges of the lattice, as shown in Fig. 5. The processing elements are being built out of conventional LSI modules (Kung, Gal-Ezer and Arun (1982)). The basic idea is that the computation starts in the top right hand corner and flows diagonally across the processors. The data fronts are "pipelined" giving rise to a series of waves flowing diagonally across the machine, which S.Y. Kung refers to as a *computational wavefront*. The key advantages that Kung, Gal-Ezer and Arun (1982) claim for the wavefront concept are:

(1) it drastically reduces the complexity of describing parallel algorithms for matrix computations,

(2) the wavefront language (developed for the machine) permits the array processor to be programmable and increases its applicability

(3) the wavefront language makes it possible to simulate and hence verify parallel algorithms, and

(4) the processors have an asynchronous waiting capability which obeys Huyghen's principle that wavefronts can never intersect.

The processor, language and applications to a large number of matrix computations are fully described in Kung, Gal-Ezer and Arun (1982) and companion papers referenced in their bibliography.

The matrix algorithms are all described for full nxn matrices using nxn processing elements and the following time-complexity results have been obtained for the machine:

(1) matrix-matrix multiplication: $nx(T_A + T_M)$

(2) LU decomposition: $nx(T_A + 2T_M + T_D)$

(3) back substitution: $nx(T_A + T_M + T_D)$

where T_A, T_M and T_D denote the time to perform addition, multiplication and division respectively.

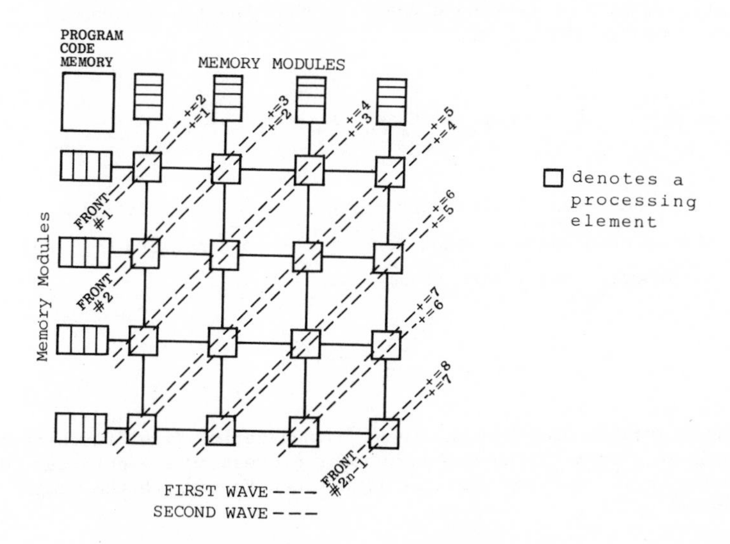

Fig. 5 The computational wave front machine
(taken from (Kung (1982c))

3.2 The Finite Element Machine

This machine has been designed, as the name would suggest, to solve the matrix equations that arise in finite element calculations. It is a multiprocessor lattice architecture with a global bus connecting all the processing elements. Because the algorithms intended for the machine are iterative in nature, the synchronisation is achieved by using special hardware which allows the multiprocessor lattice to perform one iteration and then inspects the result in each processing element's memory to test for convergence. (In the case of the conjugate gradient method, (Podsiadlo and Jordan (1981)), the hardware also computes the necessary scalar products.) The design of suitable hardware for this task is discussed in Jordan and Sawyer (1979).

The machine is designed for *sparse matrix computation* and the trick is to organise the computation so that the processing elements communicate locally and make only minimal demands on the global bus.

After the *start up phase* where each processing element receives its program and associated data, there are two principal phases to the computation; the multiplication phase and the convergence testing phase. These will be considered separately.

Multiplication Phase

In this phase the vector $\underline{r} = M\underline{x} + \underline{c}$ is computed on the lattice (the matrix M and the vector $\underline{c}$ depend on the particular iterative method being used, e.g. for the conjugate gradient method $M = A$, $\underline{c} = -\underline{b}$ and $\underline{r}$ is the residual vector). Suppose that one row of the matrix is allocated to each processing element so that the ith processing element computes the expression

$$r_i = \sum_{j=1}^{n} m_{ij}\, x_j + c_i$$

This can be achieved by first assigning $r_i = c_i$, then fetching (in turn) the value of each x_j that corresponds to a nonzero m_{ij}, and finally computing the *inner-product step*

$$r_i := r_i + m_{ij}\, x_j$$

The inner product step and the local communication (i.e. fetching the elements of $\underline{x}$ from the nearest neighbour processing element) can be performed in parallel so that the time taken to compute the vector $\underline{r}$ is bounded by

$$M_1 T_g + \lambda T_s + T_L$$

where M_1 is the number of data transfers using the global bus, T_g the time taken to move one element along the global bus, T_L time for local communication and memory fetches within the local processor's memory, T_s time to perform the inner product step and λ is the maximum number of elements in any row of A. In contrast, a uniprocessor, requires a total of $2\lambda n$ data transfers from memory to its arithmetic unit, and the total time is bounded by

$$\lambda n(T_s + T_L)$$

Thus we see that a speed-up of $O(n)$ compared with a uniprocessor is possible, provided that the number of global bus transfers, M_1 is independent of n.

<u>Convergence Testing - Inner Product Hardware</u>

Special-purpose hardware is used to test for convergence by inspecting the elements of the residual vector $\underline{r}$ and, in the case of the conjugate gradient method, to compute the necessary scalar products. If the tree-like structure, proposed in Jordan, Scalabrin and Calvert (1979) is used, then a tree of depth $\log_2 n$ is required, so the speed-up compared with a uniprocessor is $O(n/\log_2 n)$.

From the analysis we see that the potential speed-up offered by the finite element machine is particularly attractive, for equations where the sparsity of the coefficient matrix is such that extensive use can be made of the local communication links but very few demands are made on the global bus. It is very difficult to estimate the contention on the global bus because it depends on the sparsity pattern of the matrix and the way the processing elements are programmed. For this reason M. Berzins wrote a simulation program for the architecture and this work is reported in Berzins, Buckley and Dew (these proceedings).

3.2.1 Simulation of Gas Transmission Networks

The application of the multiprocessor lattice, to the simulation of gas transmission networks is considered in Dew, Buckley and Berzins (1983). The flow of gas in a steady-state gas transmission network is modelled by a sparse non-linear system of equations in which each of the unknowns represents either a gas flow or a gas pressure. The equation for the pressure at a network node depends on the gas flow along the adjoining pipes and on the pressure at the end of these pipes. Similarly the equation for the gas flow depends on the pressure at the network nodes. The sparsity pattern of the coefficient matrix in the linearised system is related to the topology of the network. Further details are given in Dew, Buckley and Berzins (1983).

Three mappings of the coefficient matrix onto the machine were investigated.

(i) Each processing element handles one solution component (e.g. the pressure at a node or the gas flow along a pipe) and one equation. The interconnection structure requires a large number of processing elements. For example a pipe requires three processing elements (one for the pipe, and two for the network nodes at each end).

(ii) Each processing element handles either a network node or a machine (e.g. a compressor or regulator). Knowing the pressures at the network nodes it is relatively easy to calculate the gas flows along the adjoining pipes.

(iii) Each processing element handles a section of the network thus reducing the number of processing elements required.

It can be seen that in moving from (i) to (iii) we have reduced the possible parallelism. The advantage of using fewer processing elements is that there is less contention for the global bus. Fortunately it is also quite easy to partition a gas transmission network to map onto a few processing elements (say up to 20 for a 100 pipe network). The network should be partitioned in such a way that ensures that roughly the same number of nodes and pipes are handled by each processing element and that the communication is, as far as is possible, via the local communication links. Because the computation time now dominates the communication time, the global bus is no longer a bottleneck in the system.

The choice of iterative method used on the finite element machine needs careful consideration. A Gauss-Seidel type iteration is often used to solve the linear equations that arise in a network simulation. When these algorithms are performed on a parallel architecture, like the finite element machine, the iteration becomes a Jacobi type. As a result the rate of convergence is likely to be slower (resulting in more iterations) or the iteration may diverge (Dew, Buckley and Berzins (1983)). For this reason it is likely to be more satisfactory to devise algorithms that use the parallel version of the conjugate gradient method described in Podsiadlo and Jordan (1981) for the finite element machine.

3.3 Systolic Array Processors using Direct Methods

In contrast to the other two machines, systolic array processors are particularly attractive for banded systems of linear equations. Consider the solution of a banded system of equations

$$A\underline{x} = \underline{b} \tag{3.1}$$

where the nxn matrix A has a band width $b_w = p+q-1$; q elements on or below the diagonal. (For example matrix A in (2.1) has p=2 and q=3.) For nonlinear problems, (3.1) is often a step in an iterative process and in this case it is necessary to solve the system of equations several times, using the solution vector $\underline{x}$ to update the matrix A and the vector $\underline{b}$.

One way of solving (3.1) is to use a direct method based on Gaussian elimination (no pivoting is considered at this stage). For this approach we need two systolic arrays (see chapter 8 of Mead and Conway (1980)).

(1) a hexagonal array of processing cells to factorise the matrix A into its LU factors (a maximum of pxq processing cells are required - see Fig. 6a), and

(2) a linear array of q processing cells to perform the back substitution.

The important point to notice is that the number of processing cells is dependent on the *bandwidth* of A and not on the number of equations, n.

In Fig. 6b, we give a possible systolic system for the solution of linear equations showing how the systolic processors can be *chained* together to ensure a smooth data flow. The system, which is a simplified version of the CMU systolic system described in section 2, illustrates the use of a controlling (or interface) processor to provide some degree of generality and to control the computation. The elements of A flow from the memory through the hexagonal array of processing cells and the LU factors from the array are stored in a stack. It takes 3n + min(p,q) units of time for all the LU factors to appear, where a unit of time is the time needed for one processing cell to get its data and compute an inner product step. Once the stack is full, it is unloaded into the back substitution systolic array with the elements of $\underline{b}$ arriving at the appropriate time (back substitution is a matrix/vector multiplication so the Kung-Leiserson algorithm can be used, see Fig. 2b, with q processing cells). The elements of the solution vector $\underline{x}$ start arriving after q units of time and the whole operation takes 2n + q units of time. The attractive feature of a systolic system is that it is not necessary to wait until a particular operation has finished. For example, as the solution elements arrive we can either start returning the results to the host or start performing any necessary updates to the matrix A and vector $\underline{b}$, and begin the solution process again. These operations

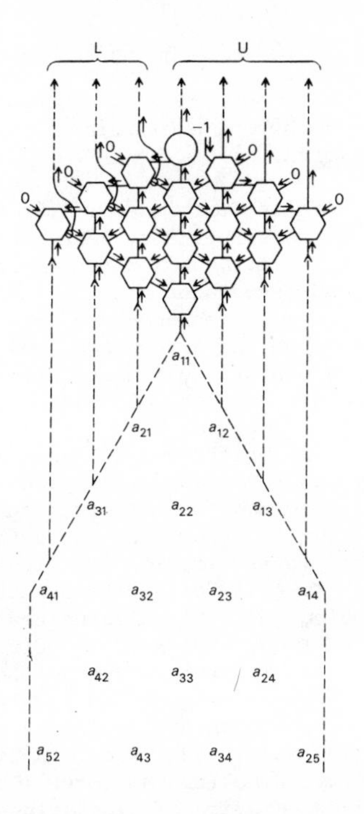

Fig. 6a The Hex-Connected Systolic Array for LU-decomposition of Matrix (2.1) taken from (Mead (1980a))

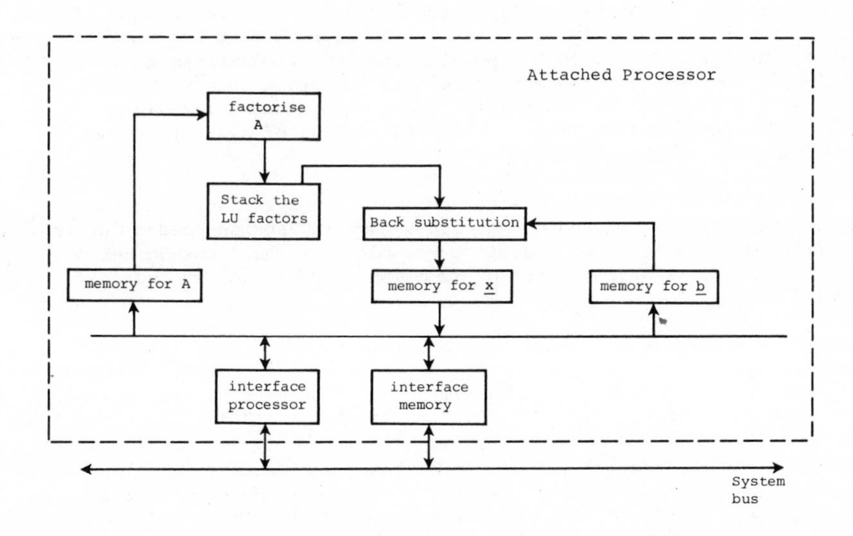

Fig. 6b A Systolic System for the Solution of $\underline{Ax}=\underline{b}$

would be under the control of the program running in the interface processor. For the solution of a large system of equations rising from partial differential equations, where the bandwidth is normally considerably less than the number of equations, a systolic system looks an attractive proposition.

Pipelined *computational arrays* for banded matrix computations have also been considered in some detail by Johnsson (1982) and Johnsson (1981). In his papers L. Johnsson considers carefully the data communication issues and introduces the idea of a *computational window* or data wavefront, see also Johnsson and Cohen (1981), Weiser and Davis (1981). The design of the arrays has been modified to handle the case when it is necessary to instantiate the computational window in both time and space.

The Kung and Leiserson LU algorithm and Johnsson's pipelined Gaussian elimination arrays do not use partial pivoting because it causes changes in the data flow and loss of performance. Instead pivoting is achieved by designing arrays for variants of the QR algorithm. A multiprocessor lattice design for Given's method was first proposed by Sameh and Kuck (1978) and later extended by Gannon (1980), for finite element problems using nested dissection and by Bojanczyka, Brent and Kung. The need to perform the square root in the Given's method has been removed in Gentleman and Kung (1981) and a systolic array for Given's method is also given in Heller and Ipsen (1982). L. Johnsson has devised a computational array for the QR-method using Householder's method Johnsson (1982) which is similar to the array in Heller and Ipsen (1982). This also provides the basis for the computation of eigenvalues and the least squares problem.

3.4 Systolic Array Processors for Matrix Iterative Methods

The systolic arrays, described above, are particularly suited to large matrices with a relatively narrow bandwidth but are less attractive when the matrix is sparse. Also we note that the stack used in the systolic system for solving the linear equations is a bottleneck because it is necessary to fill the stack before the next stage of the computation can start. These observations led Dew, Buckley and Berzins (1983) to consider the design of systolic array algorithms for matrix iterative methods.

For simplicity, we shall design and analyse a systolic array for the Jacobi iterative method using a non-symmetric, sparse-banded matrix. The modifications to other iterative methods like the Gauss-Seidel method with a relaxation factor ω, is given in Dew, Buckley and Berzins (1983).

The Jacobi iterative method can be written as

$$D\underline{x}^{(i+1)} = -(L + U)\underline{x}^{(i)} + \underline{b}, \quad i = 0,1.... \tag{5.3}$$

where L and U are respectively strictly lower and upper triangular matrices and D is the diagonal matrix, of the coefficient matrix A. The *modified Kung-Leiserson* systolic array shown in Fig. 7, for matrix (2.1), can be used to compute one Jacobi iteration. The column

CELL I - inner product step processor (figure 2a)

CELL II

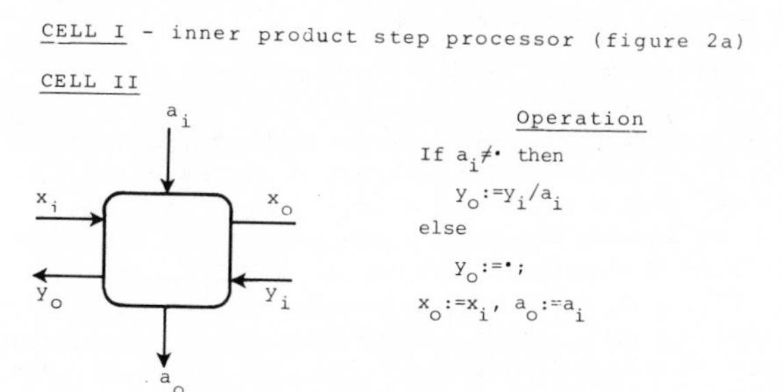

Systolic Array (modified Kung-Leiserson algorithm) for matrix (2-1)

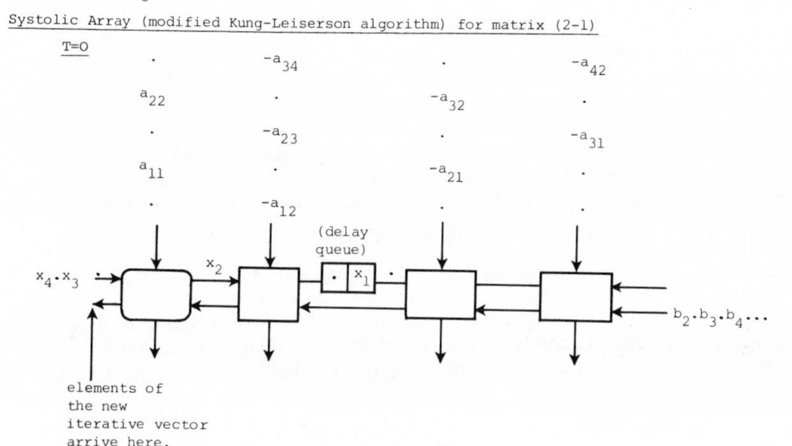

Fig. 7 A Systolic Array for the Jacobi Iterative Method

containing the diagonal element has been moved to the extreme left hand side where a special cell is used to perform the division. (Of course, this division can be avoided if we ensure that the diagonal of the matrix A is unity but this may not always be convenient or desirable.) The diagonal row of the matrix L+U is zero so we can replace one of the cells in the Kung-Leiserson systolic array by a dummy cell (Mead and Conway (1980)). However, in practice we have found that it is more convenient to introduce a delay queue in the x-stream and modify the order of the elements of A flowing into the top of the array. From Fig. 7, we see that the elements of the vector $\underline{x}^{(i)}$ flow into the left hand side of the array and the elements of the vector $\underline{b}$ flow in the right hand side. Once the pipeline has been fully loaded the elements of $\underline{x}^{(i+1)}$ arrive from the left hand side.

To obtain a significant speed-up we must pipeline several iterations. This can easily be achieved by arranging the arrays vertically so that the elements of the matrix A flow out of the bottom of one array and into next one. As the elements of the new iterative vector, $\underline{x}^{(i)}$ say, appear they feed into the systolic array immediately below so that we can start computing $\underline{x}^{(i+1)}$. To obtain the correct synchronisation it is necessary to delay the elements of A and the vector $\underline{b}$ by 2p units until the appropriate values of $\underline{x}^{(i+1)}$ have arrived.

The type of matrix that commonly occurs in a finite difference discretisation of partial differential equations has a series of nonzero bands which we refer to as a sparse-banded matrix. Such matrices arise, for example, in the finite difference discretisation of Laplace's equation in three dimensions. To handle matrices of this type, we need extend the use of delay queues between the processing cells. A suitable set of cells is shown in Fig. 8a. Cell I performs the normal inner product step but delays the output of the result by 2μ units, and cells I and II delay the matrix coefficients by 2ν units. Similarly cell III delays the elements of $\underline{b}$ by 2ν. The parameters ν and μ are constants that can be set-up at the start of the iteration giving greater flexibility. The systolic array for computing m iterations in parallel is shown in Fig. 8b.

To calculate the time complexity, suppose that the matrix A has p^* non-zero bands above the diagonal and q^* below it (both p^* and q^* do not include the diagonal). Further suppose that the element a_{1r} is the start of the first non-zero band after the diagonal element a_{11}. The vertical delay is unchanged and is equal to 2p (i.e. ν=p). It takes $\max\{2p-2p^*, q^*+1\}$ to initialise the first row (see Fig. 9) and a further p^* units for the first element of the new iterative vector to appear. It then takes a further $2p-p^*$ units to initialise the next row. Thus the time taken to compute m iterations is given by

$$\max(2p - p^*, q^* + 1) + (2p+1)(m-1) + p^* + 2(n-1)$$

units. From this result it follows that once the pipeline is full the elements of the mth iterative vector arrive at constant time.

To illustrate these results on an actual example, consider the coefficient matrix formed by a finite difference discretisation of the 2D Laplace equation within a unit square. For this problem $p=q = n^{1/2} + 1$, $p^*=q^* = 2$. The total amount of storage required by the delay queues is $O(n^{1/2})$ and the total time is $2n + mn^{1/2} + O(1)$ units.

The exact size of m is the normal trade off between hardware costs and computation speed, but ideally it should be large enough to ensure that one pass through the systolic array is sufficient. Incorporating a test for convergence is a likely bottleneck in the system because we must test all the elements of the solution vector. This could be avoided if we could estimate a bound on the number of iterations required a priori.

The introduction of delay queues is a neat solution for sparse-banded matrices but is less satisfactory for more general sparse matrices. To generalise the method we would need to re-configure the delay queue dynamically which can be achieved by sending control signals to the delay queues. These control signals essentially contain the row position of the element relative to the diagonal element.

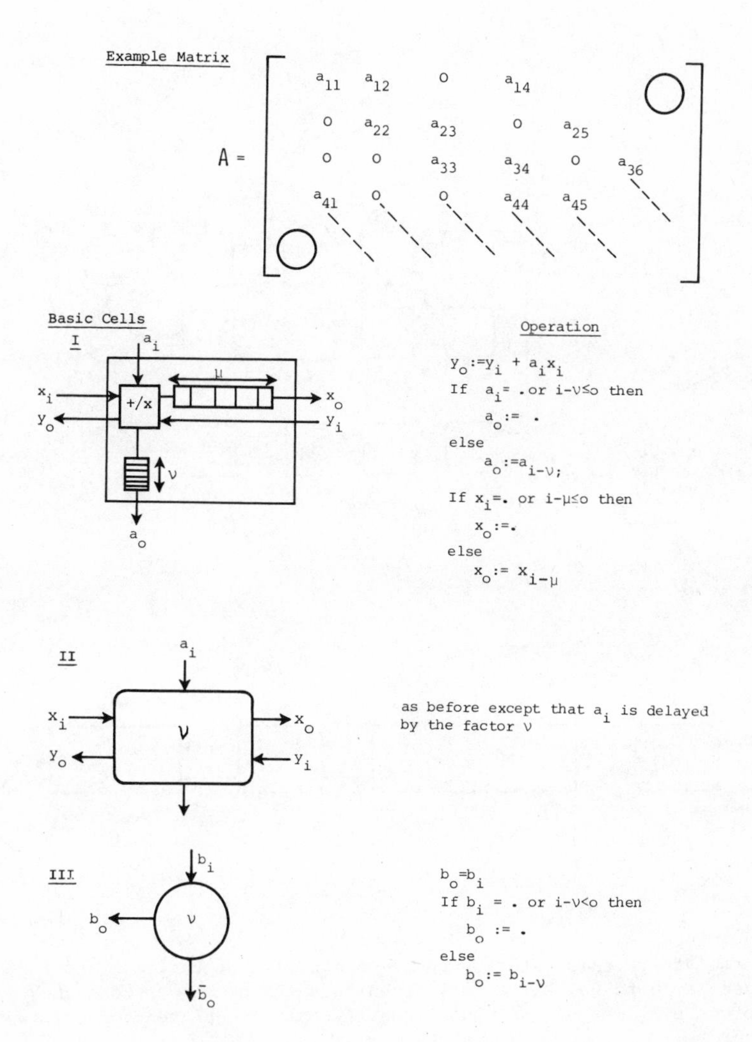

Fig. 8a Systolic array cells for Jacobi iteration

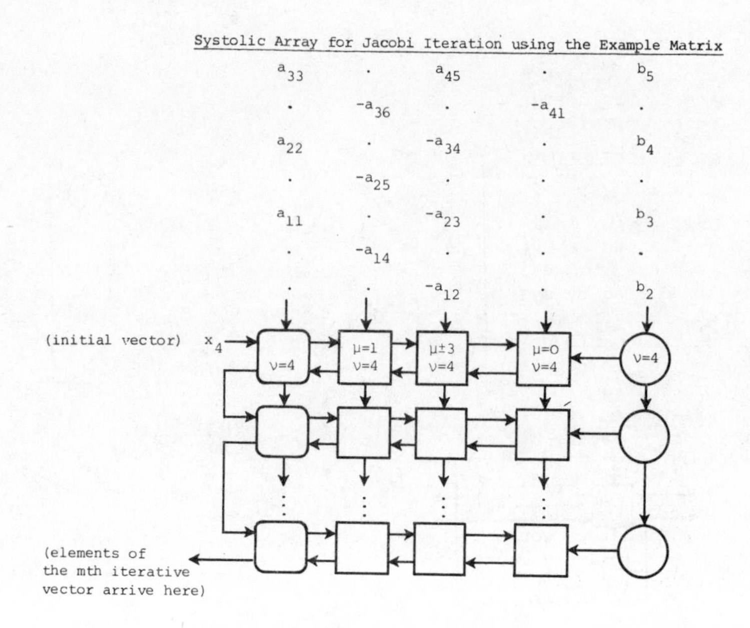

Fig. 8b Systolic Array for Jacobi iteration

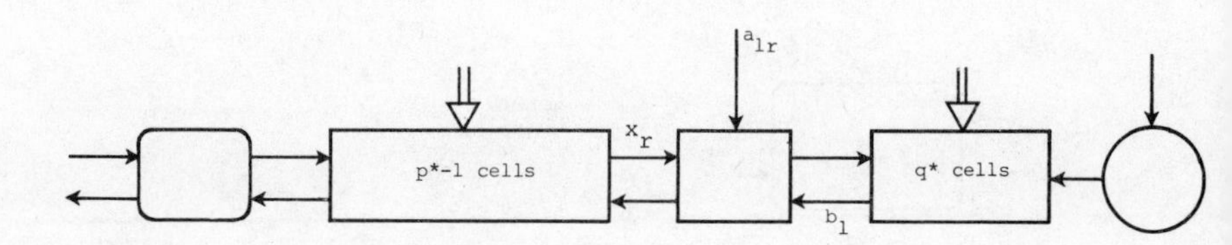

Fig. 9 The initialisation of a row of the Jacobi systolic Array

Unfortunately this interrupts the smooth flow of the data and it may be desirable to use extra cells even though a number of cells may only be processing zeros. We would identify this as an area for future research.

Another area for further investigation is the size of the bandwidth between the processing cells. It is shown in Dew, Buckley and Berzins (1983) that 16-bit precision, using fixed point arithmetic (6 places before the decimal point), was sufficient for a problem arising from the steady-state model of a water reservoir. However, it is not clear that 16-bit precision would be sufficient for many applications in numerical computation.

4. CONCLUSIONS

In this paper we have outlined some of the more interesting developments in the design of specialised parallel processors for problems in numerical computation. We have identified two broad classes of architectures, a multiprocessor lattice and systolic arrays, and considered how they can be used to solve linear equations. Of the two, the systolic array architecture looks the more promising but there is still a lot of basic research required before a general systolic system can be built. The main difficulty is likely to be building a general computing system that can provide the very wide bandwidths required by systolic array processors. Provided that these problems can be overcome, then a systolic system, like the one being designed at CMU, would be very general and it is "simply" a question of attaching the appropriate systolic arrays to perform a particular range of computations. A systolic system looks ideal for real-time control problems as we illustrated in section 2. Finally, it is worth remarking that systolic systems have much wider applicability than the general area of numerical computation.

The design of suitable systolic arrays is a fruitful area for research. One area we would identify is the development of systolic arrays for sparse matrix computations. This is important if they are going to be used to solve systems of partial differential equations.

5. ACKNOWLEDGEMENTS

The author would like to thank Martin Berzins and Tom Buckley for their helpful contributions to the author's understanding of the literature on parallel processing. He would like to thank the British Gas Corporation for supporting Martin Berzins to carry out a feasibility study into the application of VLSI to problems in the gas industry. Thanks also to J.D. Pryce and D.J. Paddon for the opportunity to present this paper.

REFERENCES

Ahmed, H.M., Delosme, J. and Morf, M. (1982) Highly Concurrent Computing Structures for Matrix Arithmetic and Signal Processing, *IEEE Computer*, **15**.

Berzins, M., Buckley, T.F. and Dew, P.M. (1982) Path Pascal Simulation of Lattice Processor Architectures for Numerical Computation, 160, Department of Computer Studies. Leeds University - see also paper in this proceedings.

Bojanczyka, A., Brent, R.P. and Kung, H.T. Numerically Stable Solution of Dense Systems of Linear Equations using Mesh-Connected Processors, to appear in *SIAM J. Scientific and Statistical Computing*.

Bokhari, S.H. (1981) On the Mapping Problem, *IEEE Trans. on Computers*, C-**30**(3), pp. 207-214.

Denyer, P.B. (1982) An Introduction to Bit-Serial Architectures for VLSI Signal Processing, in Lecture Notes for CREST/ITG Advanced Course on VLSI architectures held at the University of Bristol, (P.C. Treleaven, ed.).

DeRuyck, D.M., Snyder, L. and Unruh, J.D. (1982) Processor Displacement: An Area-Time Tradeoff Method for VLSI Design, in Proc. of MIT Conf. on Advanced Research in VLSI. (P. Penfield Jr, ed.), Artech House, pp. 182-187.

Dew, P.M., Buckley, T.F. and Berzins, M. (1983) Applications of VLSI Devices to Computational Problems in the Gas Industry, 163, Department of Computer Studies, Leeds University.

Eisenstat, S.C. and Schultz, M.H. (1981) On Some Trends in Elliptic Problem Solvers, in Proc. of Conf. on Elliptic Problem Solvers, (M.H. Schultz, ed.), Academic Press, pp. 99-114.

Finney, J. (1983) A Study of Self Tuners and Electrohydraulic Cylinder Drives. Ph.D. Thesis, Department of Mech. Eng., Leeds University.

Gannon, D.B. (1980) On Pipelining a Mesh-Connected Multiprocessor for Finite Element Problem by Nested Dissection, in Proc. Conf. on Parallel Processing, pp. 197-204.

Gentleman, W.M. and Kung, H.T. (1981) Matrix Triangularization of Systolic Arrays, Proc. of SPIE symp. : Real-Time Signal Processing IV, **298**.

Haynes, L.S., Lau, R.L., Siewiorek, D.P. and Mizell, D.W. (1982) A Survey of Highly Parallel Computing, *IEEE Computer*, **15**, pp. 9-35.

Heller, D.E. and Ipsen, I.C.F. (1982) Systolic Networks for Orthogonal Equivalence Transformations and their Applications, in Proc. MIT Conf. on Advanced Research in VLSI. (P. Penfield Jr., ed.), Artech House, pp. 130-135.

Ibbett, R.N. (1982) The Architecture of High Performance Computers, Macmillan Press.

Johnsson, L. (1981) Computational Arrays for Band Matrix Equations, 4287, Department of Computer Studies, CALTECH.

Johnsson, L. (1982) Pipelined Linear Equation Solvers and VLSI 5003, Department of Computer Science, CALTECH.

Johnsson, L. (1982) A Computational Array for the QR-method, in Proc. MIT Conf. on Advanced Research in VLSI. (P. Penfield Jr., ed.), Artech House, pp. 130-135.

Johnsson, L. and Cohen, D. (1981) A Mathematical Approach to Modelling the Flow of Data and Control in Computational Networks, in CMU Conf. on VLSI Systems and Computations. (H.T. Kung, ed.), Computer Science Press, pp. 213-225.

Jordan, H.F., Scalabrin, M. and Calvert, W. (1979) A Comparison of Three Types of Multi-processor Algorithms, in Proc. Conf. Parallel Processing, pp. 231-238.

Jordan, H.F. and Sawyer, P.L. (1979) A Multi-Microprocessor System for Finite Element Structural Analysis, *Computers and Structures,* **10**, pp. 21-29.

Kung, H.T. (1979) Let's Design Algorithms for VLSI Systems, in Proc. of Conf. on VLSI : Architecture Design Fabrication CALTECH, pp. 65-90.

Kung, H.T. (1982) Why Systolic Arrays? *IEEE Computer,* **15**, pp.37-46.

Kung, H.T. (1982) in Lecture Notes for CREST/ITG Advanced Course on VLSI Architecture, held at the University of Bristol, (P.C. Treleaven, ed.).

Kung, S.Y., Gal-Ezer, R.J. and Arun, K.S. (1982) Wave-front Array Processor : Architecture, Language and Applications, in Proc. of MIT Conf. on Advanced Research in VLSI, (P. Penfield Jr., ed.), Artech House, pp. 4-19.

Leiserson, C.E. and Saxe, J.B. (1981) Optimizing Synchronous Systems, in Proc. 22nd Symp. on Foundations of Computer Science, IEEE Computer Society, pp. 23-36.

Love, H.H. (1982) Reconfigurable Parallel Array Systems, in Designing and Programming Modern Computers and Systems, Volume **1**, LSI Modular Computer Systems, (S.P. Kartashev, et al., eds.), Prentice Hall.

Mead, C. and Conway, L. (1980) Introduction to VLSI systems, Addison-Wesley.

Pawley, S. and Thomas, G. Physical Review Letters, **48**, p. 41 (see also *New Scientist*, **1** April 1982, p. 20).

Peterka, V. (1975) A Square Root Filter for Real Time Multivariate Regression, *Kybernetika,* **11**, pp. 53-67.

Podsiadlo, D.A. and Jordan, H.F. (1981) Operating System Support for the Finite Element Machine, in Proc. of CONPAR, Lecture Notes in Computer Science, No. 111, Springer Verlag.

Ridgway-Scott, L. (1981) On the Choice of Discretisation for Solving P.D.E.s on a Multi-Processor, in Proc. of Conf. on Elliptic Problem Solvers, (M.H. Schultz, ed.), Academic Press, pp. 419-422.

Rogers, M.H. (1982) Specification of Algorithms for Systolic Array Elements, in Lecture Notes for CREST/ITG Advanced Course on VLSI Architecture, held at the University of Bristol, (P.C. Treleaven, ed.).

Sameh, A.H. and Kuck, D.J. (1978) On Stable Parallel Linear System Solvers, *ACM,* **25**, pp. 81-91.

Seitz, C.L. (1982) Ensemble Architectures for VLSI: A Survey and Taxonomy, in Proc. of MIT Conf. on Advanced Research in VLSI, (P. Penfield Jr., ed.), Artech House, pp. 130-135.

Snyder, L. (1982) Introduction to the Configurable Highly Parallel Computer, *IEEE Computer*, **15**, pp. 47-56.

Weiser, U. and Davis, A. (1981) A Wavefront Notation Tool for VLSI Array Design, in CMU Conference on VLSI Systems and Computations, (H.T. Kung et. al., ed.), Computer Science Press, pp. 226-234.

PATH PASCAL SIMULATION OF MULTIPROCESSOR LATTICE ARCHITECTURES FOR NUMERICAL COMPUTATIONS

M. Berzins, T.F. Buckley and P.M. Dew

(Department of Computer Studies, The University, Leeds)

ABSTRACT

This paper describes the simulation of multiprocessor architectures based on a lattice of processors. The general features of these architectures are discussed and the Path Pascal simulation language is briefly described. Simple Path Pascal models of the architectural features are introduced and used to investigate the efficiency of possible multiprocessor lattice architectures.

1. MULTIPROCESSOR LATTICE ARCHITECTURES

Recently several designs have been proposed for the construction of general purpose parallel architectures based on lattices of processors (see Dew (1982) for a survey). The main aim behind these designs is to take advantage of the recent advances made in VLSI technology. This should, in the near future, make it possible to construct powerful and comparatively inexpensive parallel processors based on large numbers of identical processors. We are interested in the applicability of such architectures to solving the sparse systems of nonlinear equations which arise in the simulation of gas transmission networks, see Dew, Buckley and Berzins (1982).

The general multiprocessor lattice considered here is essentially the Finite Element Machine (Podsiadlo and Jordan (1981)) and consists of the following features.

1. A host processor which transfers data and programs to and from the lattice of processors. These processors execute concurrently but interact with each other through the following three mechanisms.
2. An interconnection network of local communications lines which are controlled by programmable switches. For instance in the Finite Element Machine each processor is connected to eight lines and so there are 4N lines in total.
3. Some form of global communications network which enables a processor to communicate with every other processor in the lattice. For the Finite Element Machine this takes the form of a global data bus whereas the CHiP architecture (Snyder (1981)) configures its own communications network so that all processor links are individually dedicated. We shall be able to model both these options.
4. Special purpose hardware is used to take a result from each of the processors and to return the sum and the maximum of all these results to all the processors. This 'sum/max' hardware is used in the Finite Element Machine in forming the inner products used in numerical methods e.g. in testing for convergence (see Jordan, Scalabrin and Calvert (1979)).

There are four main potential sources of bottlenecks in this type of architecture. Firstly delays may be caused by the transfer of data and programs to the processors before the computation can commence. A similar problem occurs in the transfer of results from the lattice at the end of the computation. Secondly delays may be caused by having to synchronize the processors in the lattice. In particular this may occur at the beginning of the calculation and after each iteration of the algorithm. There may also be delays caused by interprocessor communications. It is important to evaluate how much time is wasted because a processor has to wait for a result from a locally or globally connected neighbour. Finally the delays associated with the sum/max hardware must be found.

The most cost effective way of investigating the effects of these delays on the multiprocessor lattice is by a comprehensive and detailed simulation. Realistic timing parameters for such simulation are provided by considering a simple hardware prototype. We shall mainly consider the simulation stage; details of the prototype are provided by Berzins, Buckley and Dew (1982).

2. SUMMARY OF PATH PASCAL

Path Pascal is a concurrent programming language which is well suited to the modelling of parallel architectures. The language has been designed as a modelling tool for systems programming and is an enhanced version of the P4 subset of Pascal. The main reason for the use of Path Pascal is the ease with which it is possible to model both the parallelism of the multiprocessor lattice architecture and its internal synchronization. For those unfamiliar with Path Pascal we shall briefly describe some of its main features and illustrate them with a simple example which is used in the simulation program. The reader is referred to Kolstad and Campbell (1980) for a more complete description.

2.1 Path Pascal Objects - Encapsulated Data

The principal aim behind the definition of Path Pascal objects is to allow only intended accesses to a given data structure. The encapsulation of the data in the Path Pascal object is achieved by using a path expression to synchronize and regulate accessing operations to the data structure. In particular the data and code contained within the object are accessible to other parts of the program only through explicitly defined entry operations. Objects are declared as extensions of the Pascal type definition and once declared they have their own storage and a copy of the objects information and synchronization constraints regarding the entry procedures. The typical form of a Path Pascal object is given by considering a hypothetical data structure with the name DATASTORE and the entry procedures ONE and TWO. The object definition takes the form:

```
   VAR DATASTORE : OBJECT
           PATH  <path expression > END;
            < Pascal declarations internal to the object >

           ENTRY PROCEDURE ONE;
                 BEGIN
                    < procedure body>
                 END;        (* OF ONE *)
           ENTRY PROCEDURE TWO;
                 BEGIN
                    < procedure body>
                 END;        (* OF TWO *)
      END;                                  (* OF DATASTORE *)
```

Access to the object is by the two entry procedures ONE and TWO preceded by a dot and the name of the object; e.g. DATASTORE.ONE or DATASTORE.TWO. Whether access is allowed is determined by the synchronization information contained in the path expression and by how many other invocations of ONE or TWO are currently being processed or are waiting to begin execution.

2.2 Path Expressions

Simple path expressions may be illustrated by referring to the object DATASTORE defined above and its entry procedures ONE and TWO. More complicated path expressions may be built up by replacing the components ONE and TWO with other path expressions from the examples below.

<path expression>	Meaning
ONE,TWO	Any number of parallel invocations of the procedures ONE and TWO are allowed.
ONE;TWO	Each invocation of procedure ONE must be followed by an invocation of TWO. There is no limit on the total number of invocations.
N:(ONE)	Only N invocations of procedure ONE may be active at any one time.
ONE;[TWO]	An invocation of ONE is necessary before TWO may be fired but once ONE has finished every invocation of TWO that is waiting will be activated simultaneously.

2.3 Path Pascal Processes

Path Pascal processes are declared in a similar way to standard Pascal procedures and are invoked in exactly the same way. Once invoked however each process has its own store and executes concurrently with the code which invoked it. Suppose that each of the processors in the lattice is modelled by a call to the process MICRO(I) where the argument I corresponds to the number assigned to the Ith processor in the lattice. The parallel execution of NPROC processors would then be invoked by:

```
FOR I:= 1 TO NPROC DO MICRO(I);
```

The synchronization of these processors is then enforced by Path Pascal objects corresponding to the local data links, the global data bus, and any spcial synchronization mechanism (such as testing for convergence globally at the end of an iteration loop).

2.4 Path Pascal Example

The following example illustrates how we may simulate one feature of the multiprocessor lattice. Consider the problem of synchronising all processors at one point in the computation. Each processor must asynchronously be able to access the synchronisation hardware (represented by a Path Pascal Object) and must stay locked in to this hardware until all the other processors are also locked in. This suggests that the synchronise object must have one asynchronous entry procedure (named LATTICE here). Each time the entry procedure LATTICE is called it must add one to the count of locked in processors. Access to the integer variable COUNT must be strictly sequential. This is enforced by an integer object nested inside SYNCHRONISE named ADDONE. Finally we need to stop each invocation of the entry procedure LATTICE from returning to its invoking process until the variable COUNT has reached the value, say, N. This is done by another internal object inside SYNCHRONISE named RETURN. Every procedure except the last one tries to run RETURN.HOLD but is blocked by the path expression of the RETURN object. The last processor (COUNT = N) runs RETURN.SETUP which in turn allows all the other processors to execture the HOLD procedure and to exit from the hardware. The final form of the complete code of the object SYNCHRONISE, for a 20 processor lattice, is given by:

```
SYNCHRONISE : OBJECT
              PATH LATTICE END; (* Asynchronous access*)
   CONST N = 20;                (* No of active proc   *)
   VAR COUNT : INTEGER;         (* Counts no of proc
                                   which have submitted *)
   ADDONE : OBJECT
            PATH 1:(TOCOUNT) END;  (* Sequential Access *)
      ENTRY PROCEDURE TOCOUNT;
            BEGIN
            COUNT:= COUNT+1
            END;                   (* of TOCOUNT *)
   END;                            (* of ADDONE  *)
   RETURN : OBJECT
            PATH SETUP;[HOLD] END;
      ENTRY PROCEDURE SETUP;
            BEGIN
            (* Reset COUNT for the next synchronisation *)
            COUNT:= O
            END;
      ENTRY PROCEDURE HOLD;
            BEGIN (* Synchronisation only *)
            END;
   END;                            (* of RETURN *)
   ENTRY PROCEDURE LATTICE
         BEGIN
         ADDONE.TOCOUNT;
         IF COUNT = N THEN RETURN.SETUP
                      ELSE RETURN.HOLD
         END;                  (* of LATTICE *)
   INIT; BEGIN COUNT:=O  END;
         (* initialise COUNT for the first call *)
END;     (* of SYNCHRONISE *)
```

In this way when the first processor reaches the statement in its internal program given by:

SYNCHRONISE. LATTICE;

it is forced to wait until all the other processors have reached the same point in their internal programs. The initial value of COUNT is set to zero by the initialisation block INIT (a reserved word) for the object SYNCHRONISE.

3. PATH PASCAL SIMULATION PROGRAM

Running the simulation option in Path Pascal means that the internal clock inside each process is only incremented by an integer amount, say T, by calling the Path Pascal function

DELAY(T);

from inside that process or from a procedure or function accessed by it. This function allows each part of the simulation program to directly model the timing of its hardware counterpart. In particular by changing the timing parameters associated with the architecture we can swiftly identify the likely bottlenecks in the design. The number of simulated time units which have elapsed inside a process is given by calling the parameterless integer function TIME from inside that process. By appropriate use of this function we can measure the amount of time spent in performing arithmetic operations as well as that spent in data transfers.

3.1 Local Data lines

The model used in the simulation program assumes a bi-directional bit serial line with an input buffer and an output buffer at each end. This is very similar to the currently available DART (Dual Asynchronous Receiver Transmitter) chip, (see Zilog (1980)). This model allows each processor to write directly to its output buffer but requires it to 'handshake' with its input buffer. The code for a local data line together with a simple example is given by Berzins, Buckley and Dew (1982).

3.2 Global Communications Bus

In practice it is necessary for each processor in the lattice to be able to communicate with every other processor and with the host computer by using some form of global bus. There are three items which define each data transmission, the data item being moved, its origin and its destination. The path expression for the global bus object is similar to that for the local data lines except that access to read from the bus is now defined by the processor number which is the destination of the transfer. A processor trying to read from the bus must match its number with the current destination of the bus before the read can take place.

3.3 Sum/max hardware

In a numerical calculation on the multiprocessor lattice we may have to take one contribution from each processor and return the weighted sum or the maximum of these contributions to all the processors. The Finite Element Machine uses special hardware as the most efficient solution to this problem, see Jordan, Scalabrin and Calvert (1979). The internal synchronization of the architecture ensures that all active processors submit a result to this hardware before any of them proceed to the next statement in their programs. In this way the calling of the hardware also may be seen as a synchronization mechanism for the architecture. Consequently the Path Pascal model of this hardware is based on the SYNCHRONISE object described above.

4. SIMULATION RESULTS

Recent work (Podsiadlo and Jordan (1981)) suggests that the global bus may slow down the Finite Element Machine by as little as ten per cent. As a preliminary test for the simulation program we shall take this a stage further and investigate the effect of different bus speeds when combined with different processors. The problem we shall use is the banded 20 by 20 system of equations used by Podsiadlo. The sparsity pattern of this matrix is shown in Fig. 1. To illustrate how the lattice may be used to solve this problem consider the system of equations:

$$A(\underline{x})\ \underline{x} = \underline{b}$$

where the matrix A with entries A[i,j] depends on the vector $\underline{x}$. The approach used by the Finite Element Machine to assign the lattice to the solution of this problem is to allot to each processor one element of the vectors $\underline{x}$ and $\underline{b}$ and the corresponding row of the matrix A. The interprocessor communications are used to evaluate the residual of the matrix equation above. Suppose that the ith processor is assigned to the ith row of the matrix A and the ith components of the vectors $\underline{x}$ and $\underline{b}$. As the ith element of the residual is defined by

$$r_i = b_i - \sum_j A[i,j]*x_j$$

the ith processor needs to read the current value of the jth component of $\underline{x}$ only if A[i,j] is non zero. Each processor therefore broadcasts its component of $\underline{x}$ to all the processors connected to it according to the topology of the matrix. A similar scheme may be used when each processor is assigned a block of a matrix. The efficiency of this approach depends on how much use can be made of the local links between the processors. In our case the example problem can be mapped onto part of a five by five square lattice of processing elements. These elements are numbered in increasing order along the rows; processor 1 is in the top left hand corner and processor 25 in the bottom right. The lattice has cyclic connections at the boundaries as is shown by (Jordan and Sawyer (1979)). The precise details of the interconnection schemes are shown in Fig. 4.

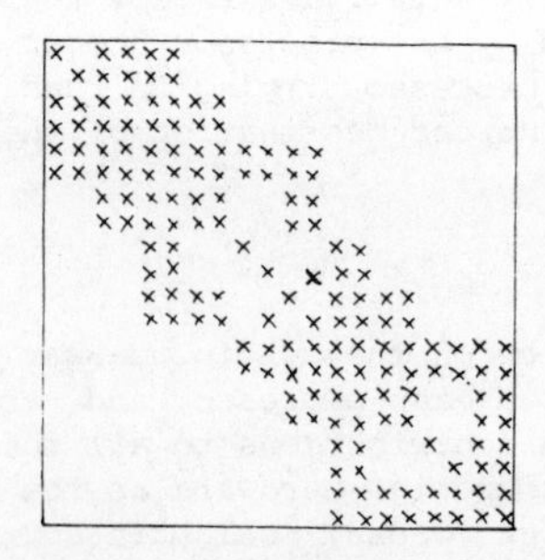

Fig. 1 Sparsity pattern of example problem

MODEL	PROCESSOR TYPE	INNER PRODUCT TIME	LOCAL COMMUNICATIONS TIME
1	Z80	65	0.65
2	Z8000	0.65	0.65
3	PURPOSE BUILT	<0.65	0.65

The inner product time is the estimated time in milliseconds to perform a = a + b x c.

Fig. 2 Details of processor types

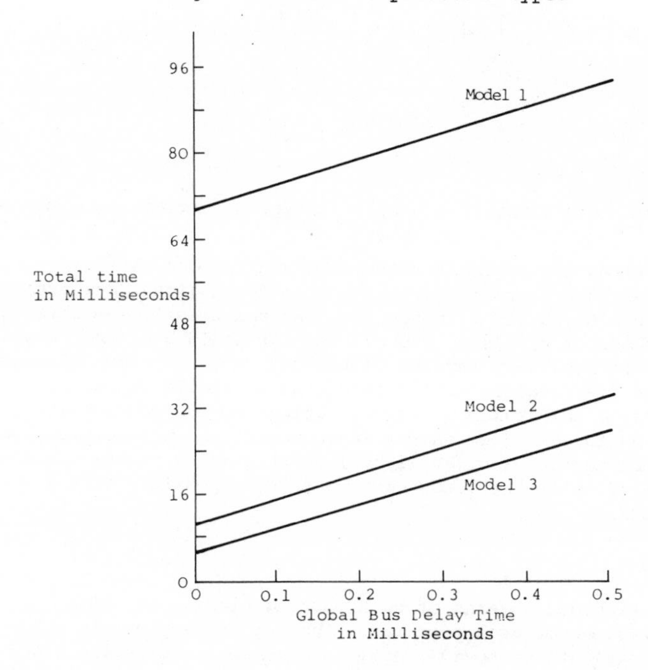

Fig. 3 Timing results for example problem

A direct mapping of the example problem consists of assigning the ith row of the matrix to processor i, in our case this means that over half the connections (44 out of 81) are made using the global bus.

In the simulation three processor designs have been considered (see Fig. 2). The first two use existing processors while the third, a purpose built processor, uses existing LSI floating point chips. The ISPS simulator (see Barbacci (1977)) was used to obtain the order of magnitude of the timing estimates employed in the simulation program. These are shown in Fig. 2. On the basis of the ISPS estimates we have assumed that it takes 325 microseconds to move a data item from the processor to the communications ports.

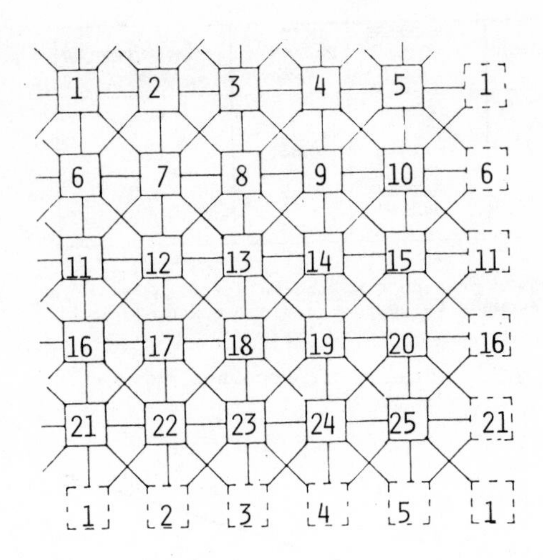

Fig. 4 The local data links in a 5 by 5 lattice of processing elements

Fig. 3 illustates the timing results obtained for one residual evaluation of the matrix problem when the time for a global bus transfer was varied between 0 and 500 microseconds. We can clearly see in all three cases the effect of a slow global bus. It would seem from these results that it is necessary for the global bus to be an order of magnitude faster than the local communications links. Should it not be feasible to achieve this ratio of communications speeds consideration ought to be given to the more general interconnections scheme proposed for the CHiP architecture (Snyder (1981)) as this would ensure a higher degree of parallelism in the communcations than is possible with a single global bus.

5. CONCLUSIONS

Path Pascal provides a means of constructing elegant but simple models of multiprocessor architectures. The use of such models makes it possible to evaluate proposed custom built architectures and algorithms in parallel. Simulation work so far indicates that it may well be necessary to have a more sophisticated communications network than the local and global links used in the Finite Element Machine.

ACKNOWLEDGEMENT

Thanks are due to Keith Hopper for introducing us to Path Pascal.

REFERENCES

Barbacci, M.R. (1977) An Architectural Research Facility, in Proc. of AFIPS 1977, pp. 161.

Berzins, M., Buckley, T.F. and Dew, P.M. (1982) Path Pascal Simulation of Multiprocessor Lattice Architectures for Numerical Computations. Dept. of Computer Studies, The University, Leeds. Report No. 160.

Dew, P.M. (1982) VLSI Architectures for Numerical Computations. These proceedings.

Dew, P.M., Buckley, T.F. and Berzins, M. (1982) The Application of VLSI Devices to Computational Problems in the Gas Industry. Dept. of Computer Studies, The University, Leeds. Report Number 163.

Jordan, H.F. and Sawyer, P.L. (1979) A Multi-microprocessor System for Finite Element Structural Analysis. *Computers and Structures*, Vol. **10**, Pergammon Press, pp. 21-29.

Jordan, H.F., Scalabrin, M. and Calvert, W. (1979) A Comparison of Three types of Multiprocessor Algorithms. Proc. 1979 International Conference on Parallel Processing, August 1979, pp. 239-248.

Kolstad, R.B. and Campbell, R.H. (1980) Path Pascal User Manual. Dept. of Computer Science, University of Illinois at Champaign-Urbana, Illinois.

Podsiadlo, D.A. and Jordan, H.F. (1981) Operating systems support for the finite element machine. CONPAR 81 (proc), Springer Verlag, Lecture Notes in Computer Science no. 111.

Snyder, L. (1981) Introduction to the Configurable Highly Parallel Computer (CHiP). Dept. of Computer Sciences, Purdue University, Technical Report CSD-TR-351.

Zilog (1980) Zilog Microcomputer Components Data Book, February 1980. Zilog Inc., California 95014.

A RECONFIGURABLE PROCESSOR ARRAY FOR VLSI

C.R. Jesshope

(Department of Electronics, University of Southampton)

ABSTRACT

Current trends in VLSI mean that a very high density of circuits can be accommodated on silicon and this in turn generates an enormously complex design problem for conventional systems architectures. Highly regular structures with regular interconnections are very necessary if large chips are to be designed with a reasonable amount of effort. One such structure with a very simple cell is the single bit processor array, the ICL DAP, Goodyear MPP and UCL CLIP are different examples of this concept. One of the possible drawbacks of these forms of architecture is the large parallelism provided, which may be more than can be used for some applications.

This paper presents a processor cell design which can be configured as a variable size array, using a trade-off between word parallelism and parallelism within the word. Chip architecture and wafer scale integration architecture will both be considered.

1. INTRODUCTION

The major problem in implementing systems in VLSI is the enormity of the design task. With current established technologies tens of thousands of gates are integrated into single designs and although there are well established methodologies for simplifying designs, this may still present many man years of effort. Future technologies are likely to make millions of gates available on a single design and unless some radical strategy is adopted the design effort will be similarly scaled.

There are two directions in which solutions to this problem can be found. One leads to more and more automation so that one will literally write hardware descriptions in some language to compile circuits onto silicon. The other direction is to design architectures or systems which minimise design effort. This can be achieved by regularity, replication or hierarchically structured designs. It must be noted that communication plays a very important part in any design for VLSI (Mead and Conway (1980)).

One computing structure which is highly regular is the processor array, an array of identical interconnected processing elements (Hockney and Jesshope (1981)). Designs for processor arrays vary considerably, with the cells of these arrays varying from single bit processors through to 64 bit floating point processors.

This paper describes a simple cell design, which the author deems to be the most cost effective end of the spectrum of designs. Several single bit processor arrays have been designed or built, including ICL DAP (Hockney and Jesshope (1981) and Flanders, et al. (1977)),

UCL CLIP (Duff (1978)), GEC GRID (Robinson and Moore (1982)) and Goodyear MPP (Hockney and Jesshope (1981)), the largest of which (MPP) will consist of 16,000 processors. Such systems, however, although ideally suited for VLSI, may have limited application due to the large parallelism exhibited. The design considered here therefore, is one in which a simple 1 bit cell may be configured (within limits) to adjust to the parallelism of a given application.

2. RECONFIGURABILITY

In a processor array there is a classic trade-off situation when considering how to utilise the hardware that may be available. This trade-off is between the power or performance of a single processor (π_∞) and the number of processors or parallelism of the system ($n_{\frac{1}{2}}$). It has been shown (Hockney and Jesshope (1981)) that the most cost effective solution to this is to choose the simplest processing element. However, this assumes that all of the parallelism produced can be utilised. A design which best meets both of these requirements is the reconfigurable array; where the point on the trade-off curve between bit and word- parallelism is not defined at system design time. Instead the choice of parallelism can be delayed until system configuration, program compilation or even program execution.

The design considered here, the Reconfigurable Processor Array (RPA), can be configured at run time and if required by the results of a user program. It consists of a very simple cell which is replicated in 2 or 3-dimensional topologies.

Examples of the effect of reconfiguration are illustrated in Fig. 1(a-d). It can be seen that on top of the basic cell structure (2-dimensions illustrated for clarity), a mapping can be performed to enable adjacent processors to be used on the same word of data. This is a logical extension to the dual processing modes available in the ICL DAP, where either bit serial processing is used or an entire row of processors may be configured to perform parallel addition. It must be noted that here, the cooperating list of cells which comprise the processing element must consist of adjacent cells in one coordinate direction in the array and may be of arbitrary length, up to the length of that dimension. Also as Fig. 1d illustrates, the configuration is not restricted to regular sets of similar sized processing elements, although these will probably be the most used configurations. It will be possible for example, to have different precision requirements in different regions of a problem being handled by processing elements of different size. This would have the effect of keeping the number of cycles for a given operation more constant.

3. CELL DESIGN AND RECONFIGURATION CONTROL

The conceptual design of the basic 1 bit cell from which processing elements are configured is given in Fig. 2. The four 1 bit state registers are as follows:

Y: accumulator

A: activity control

R: reconfiguration control

C: carry bit.

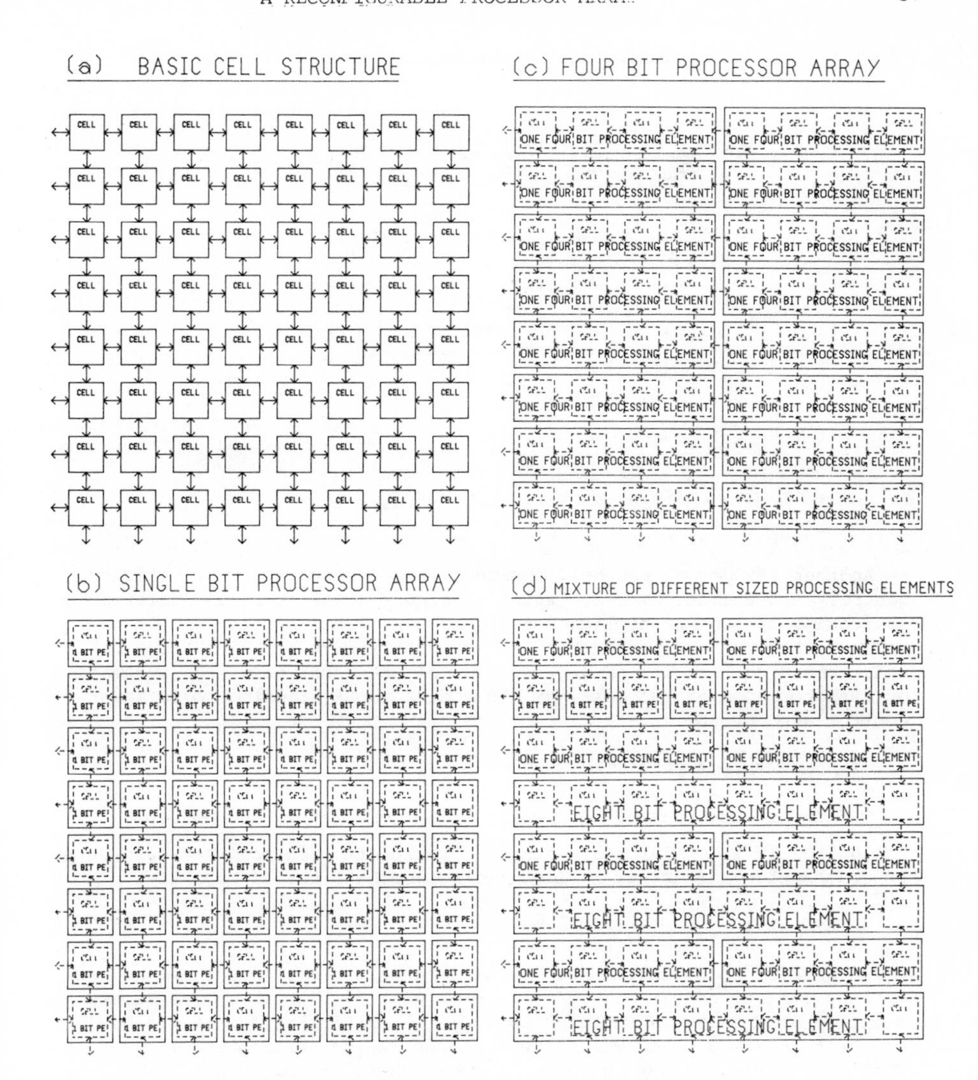

Fig. 1 Effects of reconfiguration

It can be seen that reconfiguration is controlled simply by multiplexing the input to the carry in bit of the full adder using the stored contents of the reconfiguration control. Thus for example, if the R bit were set the input would be selected from the C register, meaning that this cell was either in bit serial mode or at the least significant processing element boundary. Alternatively if the R bit were not set then the cell would be an interior bit of a multibit processing element. Thus a carry in bit is stored in the least significant cell of a processing element and a carry rippled along the length of the cell with carry propagation terminated by the next least significant or boundary cell.

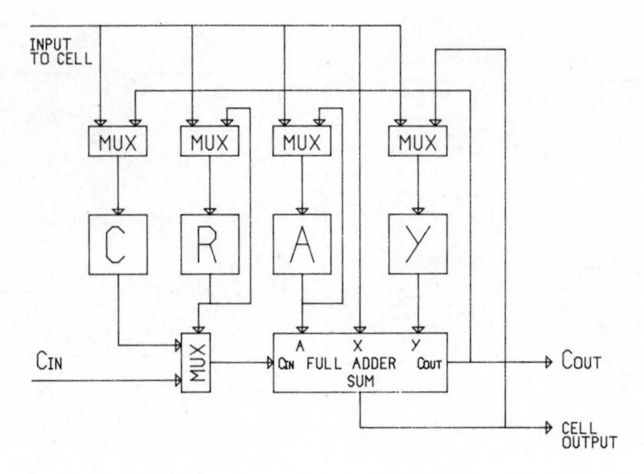

Fig. 2 Basic cell structure

Another feature of the cell design is the incorporation of activity control on the adder circuit, which is required for efficient multibit processing element multiplication algorithms. This has the effect of accumulating the carry register (using a carry save algorithm) but not the current cell input. Fig. 2 should only be considered as a schematic of the operation of the cell, as other data paths, routing logic and control are not shown.

4. IMPLEMENTATION PREDICTIONS

Two implementations are currently being considered, an NMOS chip design and an I^2L wafer scale design. These are both processes supported at Southampton University's Microelectronics Fabrication Facility.

4.1 NMOS chip

In NMOS it is estimated that a standard chip frame (about 3mm square) will be able to contain 16-32 cells each with 64 bits of random access memory. The memory is included to speed up multiplication by minimising off chip accesses which will slow down the design considerably. Preliminary simulations indicate a 100-200nsec on-chip cycle time.

4.2 I^2L wafer scale design

Array systems as described here will consist of many identical chips interconnected in a regular manner. Much of the cost of producing such a system is in the packaging, both of the chips and also boards, racks, etc. To eliminate this cost such systems could be built entirely on a single silicon wafer. The obvious problem with this is that for reasonable sized cells percentage yields on a wafer are measured in single figures.

There exist techniques and algorithms for fault detection on dynamic configurations of linear or 2-dimensional topologies (Manning (1977)). However, these rely on high yielding cell designs, typically 60%-90%.

In order to achieve these yields techniques must also be used to build fault tolerance into the cell design.

A technique to achieve this is in the use of redundancy combined with polling circuits to obtain a majority decision from the identical cells. This is illustrated in Fig. 3. It can be seen that the basic cell now has 3 inputs, 3 outputs, 3 unit cells and 3 polling circuits. This configuration can cope with failure in any one cell and polling circuit. It should be remembered that this technique could be used at many levels in the system design, from gate level upwards.

It is estimated that with redundancy to increase yield and with dynamic configuration to 'build' the array, then 4 thousand processors could be configured on a single wafer, with a reasonable amount of RAM at each cell.

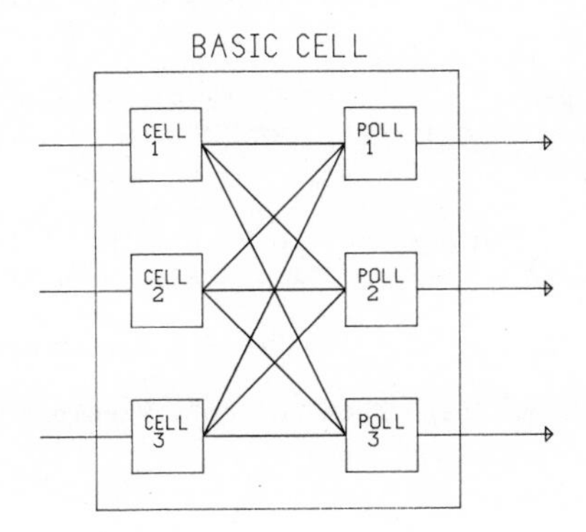

Fig. 3 Triple redundancy for wafer scale integration

5. CURRENT STATUS OF THIS PROJECT

This work has been funded under an SERC pump priming contract (GR/C 00255) under which mask making for prototype designs was awarded. At the time of writing an I^2L circuit of the basic concept in this reconfigurable design is awaiting fabrication and an NMOS design of a full processor chip with 64 bits of RAM per processor is currently being laid out.

Further funding has been requested from SERC in collaboration with Dr. W.R. Moore to follow up some of the wafer scale ideas presented above. It is anticipated that within three years a prototype chip system will be complete and a prototype wafer design will be in the process of evaluation.

Although working wafer scale designs have been fabricated there remain many problems to be overcome. These include the usefulness of the system's architecture, currently working systems have implemented linear associative machines (Wilkinson (1982) and Catt (1981)) and perhaps most importantly the techniques for controlling these architectures. Large delays will be experienced on the long runs of conductor that will be found in wafer designs, so it is not at all clear that SIMD control will be viable for wafer scale architectures.

6. REFERENCES

Mead, C. and Conway, L. (1980) "Introduction to VLSI Systems", Addison-Wesley Publishing Co.

Hockney, R.W. and Jesshope, C.R. (1981) "Parallel Computers", Adam Hilger Ltd., Bristol.

Flanders, P.M. et al. (1977) Efficient high speed computing with the distributed array processor, "High Speed Computer and Algorithm Organisation", Academic Press, pp. 113-128.

Duff, M.J.B. (1978) Clip4, "Image Processing Computers", Springer-Verlag.

Robinson, I.N. and Moore, W.R. (1982) "A Parallel Processor Array and Its Implementation in Silicon", IEEE Custom Integrated Circuit Conf, Rochester, New York.

Manning, F.B. (1977) "An Approach to Highly Integrated Computer Maintained Callular Arrays", *IEEE Trans.*, C-26, pp. 536-552.

Wilkinson, J.M., Burroughs Ltd. (1982) Private Communication.

Catt, I. (1981) "Wafer Scale Integration", Wireless World, July 1981, pp. 57-59.

EXAMPLES OF ARRAY PROCESSING IN THE NEXT FORTRAN

A. Wilson

(International Computers Limited, London)

1. The array intrinsic functions described in Section 2.9 of Brian Smith's paper (Smith, 1981) Array Processing Features in the Next Fortran fall into five groups:

- 5 functions for the measurement of arrays
- 14 functions for computation on arrays
- 2 functions for manipulation of arrays
- 3 functions for accessing arrays
- 6 functions for the construction of arrays

The names of these 30 functions (and a breakdown of the uses of the computational functions) are:

measurement: RANK, SIZE, EXTENT, LBOUND, UBOUND

computation:

- counting: COUNT
- arithmetic: SUM, PRODUCT, MASK_SUM, MASK_PRODUCT
- logical: ANY, ALL
- extremal: MAXVAL, MINVAL, MASK_MAXVAL, MASK_MINVAL
- algebraic: MATMUL, DOTPRODUCT, TRANSPOSE

manipulation: CSHIFT, EOSHIFT

access: FIRSTLOC, LASTLOC, PROJECT

construction: SEQ, ALT, DIAGONAL, SPREAD, REPLICATE, MERGE

2. The very general concepts of measurement, computation, manipulation, access and construction correctly suggest that the array intrinsic functions are intended to provide operations on arrays that are basic to the whole range of applications which use the Fortran array as their fundamental data structure. Only the small selection of algebraic computational functions MATMUL, DOTPRODUCT, TRANSPOSE is at all specialized.

In other words these functions arise naturally from the basic aim of array processing which is to operate on the arrays themselves so far as possible rather than on their individual elements.

Thus it would not achieve the aim of array processing if it was necessary to fall back on the use of loops and conditional code at every turn in order to compute such basic things as the scalar sum of the elements of a vector (= SUM(V)) or to determine the vector of the smallest positive numbers in each row of a matrix A (= MASK_MINVAL(A,A.GT.0.0,2)).

3. Here are some simple examples of array processing FORmula TRANslation:

(assume: REAL X(N), Y(N), A(M,N))

$$\sum_{j=1}^{N} \prod_{i=1}^{M} a_{ij} = \text{SUM (PRODUCT(A,1))}$$

(dimension = 1 means the product is down the columns)

$$\sum_{x_i \neq 0} 1/x_i = \text{MASK\_SUM(1/X,X.NE.0)}$$

$$\sum_{i=1}^{N} (x_i - \bar{x})^2 = \text{SUM((X-SUM(X)/N)**2)}$$

(assume: COMPLEX C(N,N))

$$v = \max_{1 \leq i \leq n} \left\{ \frac{\sum_j |c_{ij}| x_j}{x_i} \right\} \quad (x_i > 0)$$

NEW = MAXVAL(MATMUL(ABS(C),X)/X)

4. The vector $R = (r_1\ r_2\ \ldots\ r_n)$ of radii of Gerschgorin's circles associated with the complex matrix C is computed as follows:

(assume: LOGICAL OMIT(N,N))

```
OMIT = .NOT.DIAGONAL(.TRUE.,N)
   R = MASK_SUM(ABS(C), OMIT, 1)
```

5. The statistic $\lambda = \sum_{ij} \frac{(t_{ij} - e_{ij})^2}{e_{ij}}$

where E is the outerproduct of the row and column sums of T(M,N) divided by $\sum_{ij} t_{ij}$ is computed as

```
     R = SUM(T, 2)
     C = SUM(T, 1)
     E = SPREAD(R, 2, N) * SPREAD(C, 1, M)/SUM(T)
LAMBDA = SUM((T - E) ** 2/E).
```

6. The intrinsic functions allow quite complicated questions about tabular data to be answered, without use of loops or conditional code. Consider for example the questions asked below about a simple tabulation of, say, test scores.

Suppose the rectangular table T(M,N) contains the test scores of M students who have taken N different tests. (TABLE is an integer matrix with entries in the range 0 to 100)

QUESTIONS

1. What are the top scores for each student? (The list of each student's highest score)

2. How many scores in the table are above average? (average over the whole table)

3. Increase those above average scores by 10%!

4. What was the <u>lowest</u> score in the above average group?

5. Was there a student <u>all</u> of whose scores were above average?

ANSWERS

```
1.  MAXVAL(T,2)
    ABOVE = T .GT. (SUM(T)/SIZE(T))
2.  COUNT (ABOVE)
3.  WHERE (ABOVE) T = 1.1 * T
4.  MASK_MINVAL (T, ABOVE)
5.  ANY (ALL(ABOVE,2)).
```

7. The following algorithm for the solution of a tridiagonal system T x = y appears in Thomas Jordan's LASL document LA-5803, where it is described in conventional FORTRAN. The version below operates on 5 vectors of length N which represent the solution X, the righthand side Y and the 3 diagonals L, D, U (lower, diagonal, upper). It is a log_base_2 algorithm that successively drives L and U out of the picture and ends with a purely diagonal backsolution. The EOSHIFT(D,I,K) function returns D shifted left (K > 0) or right (K < 0) along dimension 1, with zero fill.

```
K = 1

DO(I = 1, LOG2N)

L = L/D

U = U/D

Y = Y/D

D = 1 - L*EOSHIFT(U, 1, -K) - U*EOSHIFT(L, 1, K)

Y = Y - L*EOSHIFT(Y, 1, -K) - U*EOSHIFT(Y, 1, K)

L =   - L*EOSHIFT(L, 1, -K)

U =   - U*EOSHIFT(U, 1, K)

K = 2* K

REPEAT

X = Y/D
```

REFERENCE

Smith, B.T. (1981) Array Processing Features in the Next Fortran, in The relationship between numerical computation and programming languages (J.K. Reid, Ed.), pp. 179-183.

OPTIMIZING THE FACR(ℓ) POISSON-SOLVER ON PARALLEL COMPUTERS

R.W. Hockney

(Computer Science Department, University of Reading)

ABSTRACT

A two parameter description of any computer is given that characterises the performance of serial, pipelined and array-like architectures. The first parameter (r_∞) is the traditional maximum performance in megaflops, and the new second parameter ($n_{1/2}$) measures the apparent parallelism of the computer. The relative performance of two algorithms on the same computer, depends only on $n_{1/2}$ and the average vector length of the algorithms. The performance of a family of FACR direct methods for solving Poisson's equation is optimized on the basis of this characterisation.

1. INTRODUCTION

The selection of algorithms for the new generation of vector and parallel processors presents special difficulties because of the wide range of different computer architectures that are represented. Some of these are described elsewhere in this symposium and also in the recent book by Hockney and Jesshope (1981) entitled "Parallel Computers". These computers range from designs with a few special-purpose pipelined functional units (such as the CRAY-1) to designs with many thousands of identical simple processors (such as the ICL DAP). Obviously it would be desirable not to have to treat each design as a special case, and in section 2 we give a two-parameter description of the hardware performance of a computer that treats serial, pipelined and array-like architectures in a uniform way. This characterisation is based on the time to perform a single vector operation as a function of vector length, and can be measured very simply on any computer. The first parameter, r_∞ is the traditional maximum performance in millions of floating-point operations per second (megaflops), and the new second parameter, $n_{1/2}$, is the vector length necessary to achieve half that maximum performance. We show that $n_{1/2}$ is a measure of the apparent parallelism of the hardware, and answers the question "How parallel is my computer?". In particular, $n_{1/2} = 0$ is the traditional serial computer, and $n_{1/2} = \infty$ is the infinitely parallel paracomputer, so beloved by analysts in complexity theory.

An algorithm may be considered as a sequence of vector instructions and, in section 3, we use the above characterisation to compare the performance of different algorithms on different computers. It is then that the significance of $n_{1/2}$ becomes apparent: its value determines the best algorithm on a particular computer. Our method of comparing algorithms using $n_{\frac{1}{2}}$ is simple, highly practical, and requires only elementary algebra. Furthermore, the substitution of the appropriate value of $n_{\frac{1}{2}}$ allows the results to be applied to all serial and parallel designs. This is in marked contrast to traditional methods in which the operations' count is used as an appropriate figure of merit for serial computers ($n_{\frac{1}{2}} = 0$), and the number of timesteps of complexity analysis is used for the paracomputer ($n_{\frac{1}{2}} = \infty$). Thus we find that the two traditional figures of merit are extreme cases in our more general analysis which uses $n_{\frac{1}{2}}$ as a rational means of interpolation between them. Furthermore, our analysis applies to real computers that have a finite and measurable value of $n_{\frac{1}{2}}$ (or apparent parallelism).

The comparison of algorithms using $n_{\frac{1}{2}}$ is applied in section 4 to a family of direct methods for solving Poisson's equation using Fourier analysis and cyclic reduction. These comparisons are not intended to be exhaustive - there are far too many published methods than can reasonably be compared in a paper of this size. They are presented to illustrate the method of comparison and the types of conclusion that may be drawn.

2. PARALLEL COMPUTERS

A two-parameter description of the performance of any computer can be obtained by fitting the best straight line to the measured time, t, to perform a single vector operation on vectors of varying length, n, (e.g. A = B*C, where A, B and C are vectors). A similar description of computer performance has been developed by Calahan, Ames and their coworkers at the University of Michigan (see (Calahan and Ames (1979)) and the references therein). Our work below differs in the definition of parameters and the use made of them. Two equivalent generic forms for the straight line define two primary and one useful secondary derived parameter:

$$t = r_{\infty}^{-1}(n + n_{\frac{1}{2}}) \tag{1}$$

where

r_{∞}: (maximum or asymptotic performance) the maximum number of elemental arithmetic operations (i.e. operations between pairs of numbers) per second, usually measured in megaflops. This occurs for infinite vector length on the generic computer.

$n_{\frac{1}{2}}$: (half-performance length) the vector length required to achieve half the maximum performance.

Alternatively, when $n < n_{\frac{1}{2}}$, the generic line may be more usefully expressed as:

$$t = \pi^{-1}(1 + n/n_{\frac{1}{2}}) \tag{2}$$

where

π: (specific performance) or performance per unit parallelism, is defined as the ratio $r_\infty/n_{\frac{1}{2}}$.

The above definitions are shown graphically in Fig. 1 where we find:

r_∞ is the inverse slope of the generic line

$n_{\frac{1}{2}}$ is its negative intercept on the n-axis

π is its inverse intercept on the t-axis

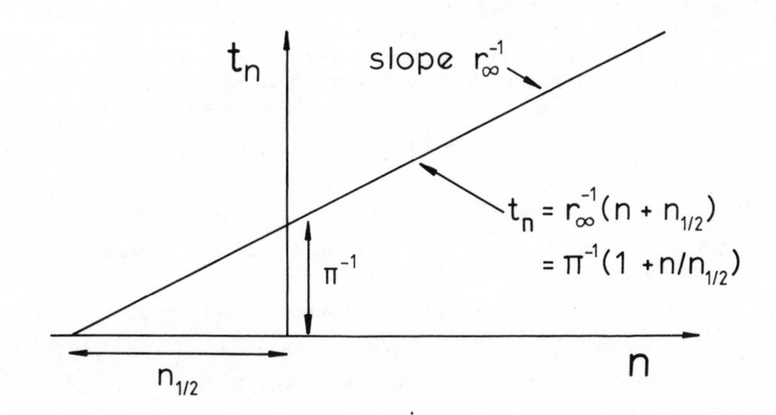

Fig. 1 The timing diagram for the generic parallel computer, showing the definitions of the parameters, r_∞, $n_{\frac{1}{2}}$ and π. (From Hockney and Jesshope (1981), courtesy of Adam Hilger.)

It is useful to examine the values of r_∞ and $n_{\frac{1}{2}}$ that are expected from the common forms of computer architecture. This is done by considering the timing line for each type:

(a) Serial Computer - the execution time is proportional to the number of elemental operations

$$t = t_1 n \tag{3}$$

where t_1 is the time for one elemental operation.

Comparison with Eqn. (1) shows that for a serial computer

$$r_\infty = t_1^{-1}, \quad n_{\frac{1}{2}} = 0 \tag{4}$$

(b) Pipelined Computer - the execution time is normally expressed by the manufacturers in a form similar to

$$t = (s + \ell + n - 1)\tau \tag{5}$$

where

τ is the clock period

s is the setup time in clock periods

ℓ is the number of segments in the arithmetic pipeline

Comparison with Eqn. (1) shows that for a pipelined computer

$$r_\infty = \tau^{-1}, \quad n_{\frac{1}{2}} = s + \ell - 1 \tag{6}$$

(c) Processor Array - if there are N processors which simultaneously perform the same arithmetic operation on N elements of each vector (one element of each vector in each processor's memory), then the timing graph is stepwise as shown in Fig. 2

$$t = t_p \lceil n/N \rceil \tag{7}$$

where

$\lceil x \rceil$ is the ceiling function of x, i.e. the smallest integer which is equal to or greater than x.

t_p is the time for one parallel arithmetic operation of all processors in the array.

The best straight line through the timing graph is the dotted line which corresponds to

$$r_\infty = N/t_p, \quad n_{\frac{1}{2}} = N/2 \tag{8}$$

This choice of parameters describes approximately the average behaviour of the array if the vectors presented to it are of varying lengths, more or less uniformly distributed.

On the other hand one may know that the vector length is always less than the number of processors ($n \leq N$) and that therefore one is always working on the first step of the timing graph. In this case the behaviour is exactly described by the second generic form with

$$\pi = t_p^{-1}, \quad n_{\frac{1}{2}} = \infty \tag{9}$$

We note that this condition is the one assumed in the complexity theory of parallel algorithms: that is to say that there are always enough processors. This can occur in general for the theoretical paracomputer which has an infinite number of processors. It is nice that in our formalism this theoretical limit occurs when $n_{\frac{1}{2}} = \infty$.

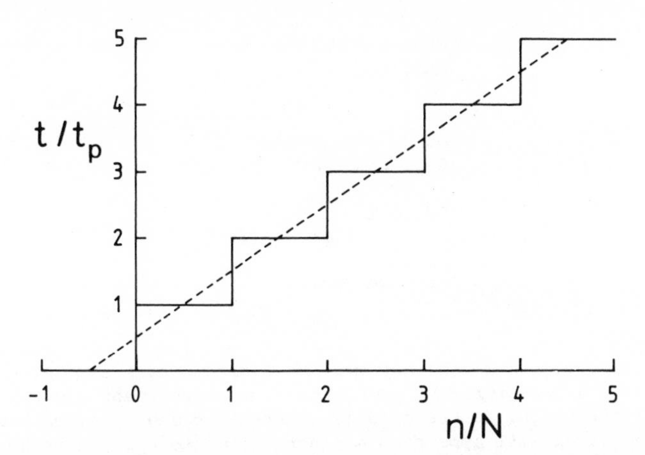

Fig. 2 The timing diagram for an array of N processing elements (solid line), showing the best approximating generic straight line (dotted) which determines the value of $n_{\frac{1}{2}}$ as N/2. (From Hockney and Jesshope 1981, courtesy of Adam Hilger.)

The above theoretical results for a range of widely different computer architectures suggest that $n_{\frac{1}{2}}$ is a measure of the parallelism of the computer hardware, varying from zero for a serial computer with no parallelism to infinity for the infinite array of processors. The exception is the pipelined computer in which a large value of $n_{\frac{1}{2}}$ can occur either for a large amount of parallel operation in the pipeline (the number of segments ℓ is large), or for a large value of the setup time, s. In the former case $n_{\frac{1}{2}}$ is measuring the hardware parallelism, but in the latter case it is measuring an overhead. From the users, or algorithmic, point of view the behaviour of the computer is determined by the timing expression (1) and the value of $n_{\frac{1}{2}}$, however it arises. A pipelined computer with a large value of $n_{\frac{1}{2}}$ appears and behaves as though it has a high level of parallelism, even though this might be due to a long setup time. Hence we regard $n_{\frac{1}{2}}$ as a measure of the apparent parallelism of the computer, and from the algorithmic (i.e. timing) point of view it simply does not matter how much of this is real. The fact that true parallelism and setup time are interchangeable, incidently, shows that parallelism is an overhead, and therefore undesirable (by which we mean that parallelism is best avoided if at all possible, or that one should always seek to achieve the required performance with the least possible parallelism).

The values of $n_{\frac{1}{2}}$ and r_{∞} of a computer are best regarded as measured quantities obtained by executing the following FORTRAN code and plotting the timing graph of T against N:

```
          CALL SECOND(T1)
          CALL SECOND(T2)
          TO = T2 - T1

          DO 20 N = 1,NMAX
          CALL SECOND(T1)

          DO 10 I = 1,N                                   (10)
10         A(I) = B(I) * C(I)

          CALL SECOND(T2)
20         T = T2 - T1 - TO
```

In the above code, the DO 10 loop will be replaced by a single vector instruction by any vectorizing compiler. The measurement and subtraction of the timing overhead TO is essential because, as we have seen, any overhead will appear as a contribution to $n_{\frac{1}{2}}$. In this case the overhead of measurement is nothing to do with the time of execution of the vector operation, and must therefore be subtracted.

The characterisation of the performance of computers by two parameters naturally leads to plotting computers as points in the two-dimensional $(n_{\frac{1}{2}}, r_{\infty})$ phase plane, as is done for some well known designs in Fig. 3. In practice most computers may operate in different modes (scalar or vector, dyadic or triadic operations, different word lengths etc.) and therefore appear as a series of dots, joined to form a "constellation" in the diagram. The traditional characterisation of computer performance by the single parameter r_{∞} corresponds to projecting this diagram onto, or viewing it through, the vertical axis. In the era of serial computers all of which have the same $n_{\frac{1}{2}}$ of zero, this was clearly valid. However in the age of the parallel computer, it is obviously important to recognise the different levels of apparent parallelism by spreading the computers out along the $n_{\frac{1}{2}}$ axis. We call this the two-dimensional spectrum of computers. We shall see in the next section that $n_{\frac{1}{2}}$ determines the choice of the best algorithm, and hence is a very important axis. As examples, Fig. 3 shows that the CRAY-1 ($n_{\frac{1}{2}} \approx 10$) and the CYBER 205 ($n_{\frac{1}{2}} \approx 100$) have similar values of r_{∞} but behave very differently because their values of $n_{\frac{1}{2}}$ differ by a factor 10. For the same reason, the ICL DAP ($n_{\frac{1}{2}} \approx 1000$) differs from both the CRAY-1 and CYBER 205.

3. PARALLEL ALGORITHMS

To a first approximation an algorithm can be regarded as a sequence of vector operations of varying length (including one). Such a representation, of course, neglects many factors that may be important (even crucial) in particular cases. Such factors may be, for example, memory bank conflicts in pipelined computers, data routing delays in processor arrays, and the simultaneous operation of scalar and vector units. However we have to start somewhere and avoid too much complication if we are to obtain manageable results. Therefore, in common with other theoretical analyses of algorithm performance, we shall assume such

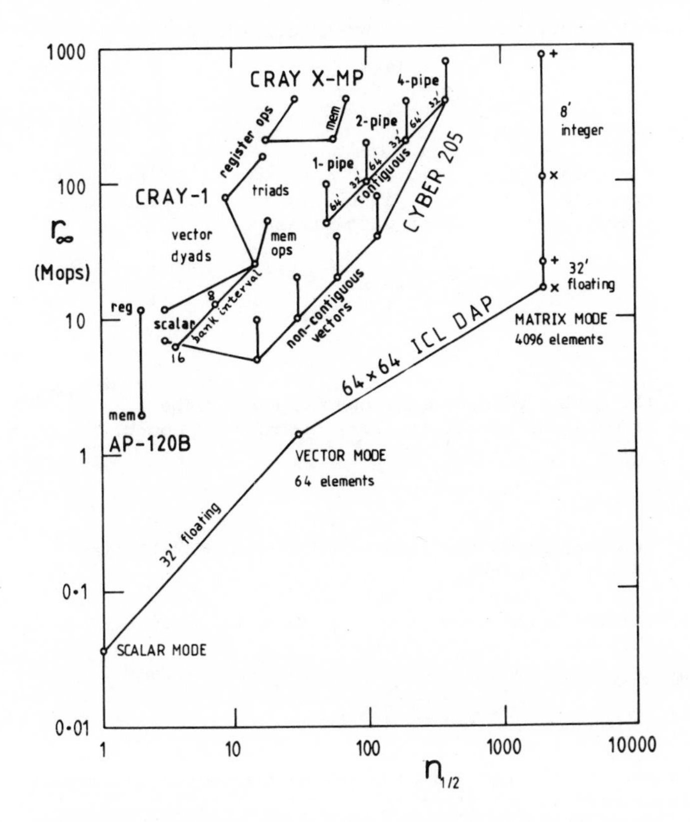

Fig. 3 The two-dimensional spectrum of computers, showing the CRAY-1, CYBER 205 and ICL DAP. (After Hockney and Jesshope 1981, courtesy of Adam Hilger.)

factors are unimportant and express the total time T, for the execution of an algorithm as

$$T = r_\infty^{-1} \sum_{\ell=1}^{\ell max} q_\ell (p_\ell + n_{\frac{1}{2}}) \tag{11}$$

where we regard the algorithm as ℓmax sequential stages, ℓ, each composed of q_ℓ vector operations of length p_ℓ. The generic timing formula (1) is then used to buildup the expression (11).

It is useful to define the following quantities:

$$q = \sum_{\ell=1}^{\ell max} q_\ell$$

the total number of vector operations, the parallel operations' count or, in the language of complexity theory, the number of unit timesteps.

$$s = \sum_{\ell=1}^{\ell max} q_\ell p_\ell$$

the number of elemental operations, or the traditional serial (scalar) operations' count.

$$\bar{p} = s/q$$

the average vector length, or average parallelism of the algorithm.

Using these variables the time of execution of an algorithm can be expressed either as

$$T = r_\infty^{-1} q(\bar{p} + n_{\frac{1}{2}}) \tag{12}$$

where the algorithm is regarded as q sequential vector operations with average vector length $\bar{p}$, or as

$$T = r_\infty^{-1} (s + n_{\frac{1}{2}} q) \tag{13}$$

where the first term is the contribution from the traditional count of all elemental arithmetic operations, and the second term is the contribution from the number of parallel (i.e. vector) operations. Equation (13) demonstrates clearly the role of $n_{\frac{1}{2}}$ in interpolating between the extremes of the serial computer ($n_{\frac{1}{2}} = 0$) and the infinitely parallel computer ($n_{\frac{1}{2}} = \infty$). For serial computers only the first term or elemental operations' count matters. For the infinitely parallel computer only the second term or the number of parallel operations matters. For computers with finite parallelism, a linear combination of the two operations' counts is appropriate, and the value of $n_{\frac{1}{2}}$ gives the weighting between the two. Since $n_{\frac{1}{2}} = \infty$ corresponds to the assumptions made in the complexity analysis of parallel algorithms and q is the number of unit timesteps in such an analysis, equation (13) shows also how $n_{\frac{1}{2}}$ interpolates rationally between the extreme assumptions that are used in complexity analysis and those that have traditionally been used in the analysis of algorithms on serial computers.

It is instructive to relate the quantities defined above to those introduced by Kuck (1978) for the analysis of parallel algorithms. The most important of these is SPEEDUP which relates the speed of an algorithm on a parallel multiprocessor array to the speed of the same algorithm on a serial uniprocessor with the same speed arithmetic units. Thus

$$\text{SPEEDUP} = \frac{\text{time of execution on uniprocessor}}{\text{time of execution on multiprocessor}} = \frac{\text{number of elementary operations}}{\text{number of parallel operations}} = \frac{s}{q} = \bar{p} \tag{14}$$

That is to say the SPEEDUP is nothing other than the average vector length (or parallelism) of the algorithm.

The use of the SPEEDUP factor as a figure of merit for parallel algorithms can be misleading because it is only one of several factors that must be considered in any comparison between a real parallel multiprocessor array and a real serial uniprocessor. Let us define the performance (or speed), P, of an algorithm as the inverse of its time of execution, that is to say T^{-1}, the number of executions of the algorithm that are possible per second. Then the relative performance is given by

$$\frac{P_p}{P_s} = \frac{T_s}{T_p} = \frac{s_s \times t_s}{q_p \times t_p} \tag{15}$$

where the subscripts s and p refer to the serial uniprocessor and parallel multiprocessor respectively, and t_s and t_p are respectively the time for a serial and a parallel operation. Equation (15) can be expressed as

$$\frac{P_p}{P_s} = \frac{s_p}{q_p} \times \frac{s_s}{s_p} \times \frac{t_s}{t_p} = \text{SPEEDUP} \times \text{algorithmic SLOWDOWN} \times \times \text{hardware SLOWDOWN} \tag{16}$$

The first factor in equation (16) is the SPEEDUP factor previously defined, however the second and third factors are SLOWDOWN factors. In order for the parallel multiprocessor to outperform the serial uniprocessor, it is necessary that the product of the SPEEDUP and the SLOWDOWN factors be greater than one. It is not sufficient that the SPEEDUP factor alone be greater than one. The first SLOWDOWN factor, the algorithmic SLOWDOWN, arises because the definition of SPEEDUP assumes that the parallel algorithm is executed on the serial uniprocessor with an elemental operations' count of s_p. Almost certainly an algorithm chosen for a parallel computer will not be the best on a serial computer, and the number of elemental operations in the best serial algorithm s_s will almost certainly be less than s_p. Hence the algorithmic SLOWDOWN factor $\frac{s_s}{s_p} < 1$ (typically 1/5).

The second SLOWDOWN factor, the hardware SLOWDOWN, expresses the fact that if the multiprocessor and uniprocessor consume comparable resources, either in money, in number of chips, or in square millimetres of silicon, then the time to perform a serial operation on the uniprocessor, t_s will be much less than the time to perform a parallel operation on the multiprocessor, t_p. In other words if you build many thousands of processors, each of them is going to be very slow compared with the speed of a single processor built or purchased with the same resources. Hardware SLOWDOWN factors are likely to be very small ($\approx 10^{-3}$ to 10^{-4}). To take an extreme example, the CRAY-1 acts like a serial uniprocessor (small $n_{\frac{1}{2}} \approx 10$) and can produce an arithmetic result every 12.5 ns ($=t_s$). On the other hand, the ICL DAP is a parallel array of 4096 processors and performs a parallel operation in about 250 μs ($=t_p$). For these two computers the hardware SLOWDOWN is about 1/20,000. Taking the two example SLOWDOWN factors, we see that the SPEEDUP might have to exceed 100,000 before the parallel multiprocessor is likely to outperform the serial uniprocessor.

Traditional methods for comparing the performance of algorithms are based either on the assumption that the computers are serial, when we compare the elemental operations' count, s; or on the assumption that the computers are array-like and always with sufficient processors, when we compare the parallel operations' count, q. We prefer to use the more general timing expression (12) or (13) and obtain a performance comparison for computers with finite values of $n_{\frac{1}{2}}$. Suppose we compare the performance of algorithm (a) on computer (1) with algorithm (b) on computer (2), then

$$\frac{P^{(a,1)}}{P^{(b,2)}} = \frac{T^{(b,2)}}{T^{(a,1)}}$$

$$= \frac{(s^{(b)} + n_{\frac{1}{2}}^{(2)} q^{(b)})}{(s^{(a)} + n_{\frac{1}{2}}^{(1)} q^{(a)})} \times \frac{r_\infty^{(1)}}{r_\infty^{(2)}} \times \frac{C^{(2)}}{C^{(1)}} \tag{17}$$

In the above, superscripts are used to distinguish the computer or algorithm; and we note that the algorithm is specified by the value of s and q (or $\bar{p}$ and q) and the computer is specified by values of $n_{\frac{1}{2}}$ and r_∞ (or $n_{\frac{1}{2}}$ and π). The first two factors in Eqn. (17) come from the timing expression (13) and the last factor may be added if the cost of computer time is a relevant factor. C denotes the cost per unit computer time.

Equation (17) is general and compares the cost performance of different algorithms on different computers. If, however, we limit consideration to the choice of the better algorithm (in the sense of having the higher performance) on a particular computer, then Eqn. (17) reduces to

$$\frac{P^{(a)}}{P^{(b)}} = \frac{s^{(b)} + n_{\frac{1}{2}}q^{(b)}}{s^{(a)} + n_{\frac{1}{2}}q^{(a)}} \tag{18}$$

in which we note that the second and third factors in Eqn. (17) reduce to unity, and that the choice of the better algorithm depends only on the $n_{\frac{1}{2}}$ of the computer and the s and q operations' counts of the algorithms.

In the comparison of algorithms, the equal performance line along which $P^{(a)} = P^{(b)}$ plays a key role because it divides regions of phase planes in which algorithm (a) has the better performance from regions in which algorithm (b) has the better performance. Along the equal performance line we have

$$n_{\frac{1}{2}} = \frac{s^{(b)} - s^{(a)}}{q^{(a)} - q^{(b)}} \tag{19}$$

the left-hand side of which depends only on the computer and the right-hand side only on the algorithm. In general the operations' counts s and q are non-linear functions of some quantity measuring the size of the problem being solved: for example the dimension, n, of the matrices in a matrix problem. The equal performance line (19) can then easily be drawn on the $(n_{\frac{1}{2}},n)$ phase plane, because $n_{\frac{1}{2}}$ is always an explicit function of n, albeit a non-linear one. The phase plane can thereby be divided into regions in which each algorithm has the better performance. Sometimes it may be desirable from the graphical point of view to scale the axes and plot, for example, the $(n_{\frac{1}{2}}/n,n)$ or $(n_{\frac{1}{2}}/n^2,n)$ phase plane. It is a useful convention, however, always to choose the x-axis proportional to $n_{\frac{1}{2}}$, the apparent parallelism of the computer. In this way serial computer algorithms always appear to the left of the diagram, and parallel computer algorithms to the right.

4. POISSON'S EQUATION

In this section we apply the method of analysis developed in section 3 to the selection of the best member of a family of direct methods for the solution of the model Poisson problem. The problem is the solution of the 5-point difference approximation to Poisson's equation on a square $n \times n$ finite difference mesh with simply boundary conditions (either given value, gradient or periodicity). Such a problem may seem artificially simple and of little practical importance, however history has shown that there are many important problems in physics (plasma, astro-, and dense matter), electrical engineering (semiconductor device simulation) and meteorology that require especially rapid methods for solving this problem (see, for example, Potter (1973); Hockney and Eastwood (1981)).

The method to be analysed is direct, and involves the optimum combination of Fourier analysis in the x-direction and block cyclic reduction by lines in the y-direction. The method is known as the FACR(ℓ) algorithm, where ℓ is the number of stages of line cyclic reduction that are performed before Fourier analysis takes place. It represents a family of algorithms because the parameter ℓ can be used to minimise the time of execution. The first algorithm in this family, FACR(1), was published in 1965 by Hockney (1975) working in collaboration with Golub. Subsequently the optimum value of $\ell(\approx\log_2\log_2 n)$ for serial computers was discovered empirically by Hockney (1970), and the asymptotic form given later by Swarztrauber (1978).

On parallel computers, it is interesting that the optimum value of ℓ depends not only on the size of the problem, n, as it does on a serial computer, but also on the parallelism of the computer as measured by its half-performance length, $n_{\frac{1}{2}}$. Hockney and Jesshope (1981) have given the analysis for one way of implementing the FACR algorithm on a parallel computer which is most suitable for low levels of parallelism (the SERIFACR algorithm). Here we extend the previous work to a way of implementation that maximises the parallelism (i.e. vector length) of the algorithm and is most suitable for highly parallel computers (the PARAFACR algorithm). The reader is referred to the above book for a derivation of the operations' counts for Fourier analysis and cyclic reduction. We will quote these here and concentrate on the problem of finding the optimum value of ℓ.

A. The SERIFACR Algorithm

The FACR algorithm involves five stages, and the variables that are related in each stage are shown for the FACR(1) algorithm in Fig. 4. The stages are:

(a) Modify RHS - block cyclic reduction by lines means the modification of the right-hand side of the Poisson equation on $n2^{-r}$ lines, where $r = 1,2,\ldots,\ell$. Vectors are run in the vertical direction, and are composed of corresponding variables in each of the $n2^{-r}$ lines. The vector length is therefore $n2^{-r}$. The number of parallel operations is $(3 \times 2^{r-1} + 2)n$, thus the time for this stage of the algorithm is proportional to

$$t_a = \sum_{r=1}^{r=\ell} (n_{\frac{1}{2}} + n2^{-r})(3 \times 2^{r-1} + 2)n \qquad (21)$$

the factor r_∞^{-1} is omitted in the above because, as was seen in section 3, it cancels out in any comparison of different algorithms on the same computer.

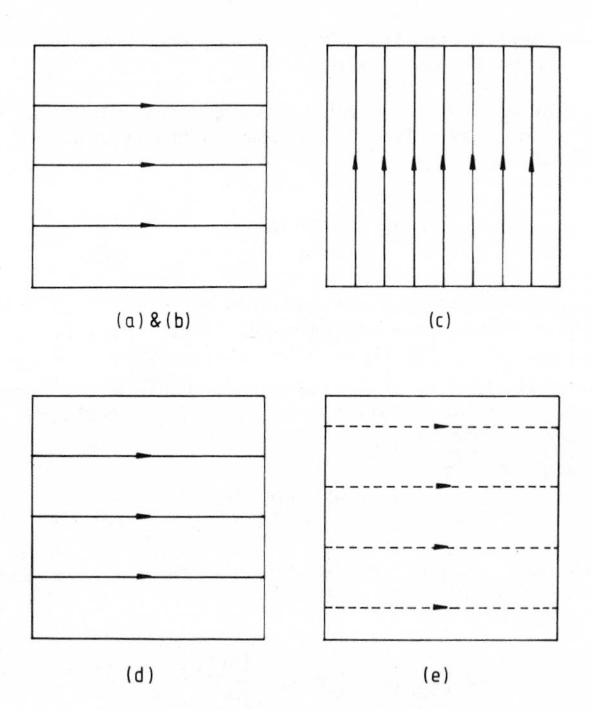

Fig. 4 Data relationships in the FACR(1) algorithm. The arrowed lines join variables related by equations or an FFT during different stages of the algorithm. (From Hockney and Jesshope 1981, courtesy of Adam Hilger.)

(b) Fourier analysis - is performed on $n2^{-\ell}$ lines in parallel. Vectors are of length $n2^{-\ell}$ and run vertically across the lines. The transforms are real and of length n and can be performed by the fast Fourier transform (FFT) in $2\frac{1}{2}n\log_2 n$ vector operations of length $n2^{-\ell}$, hence

$$t_b = (n_{\frac{1}{2}} + n2^{-\ell})\,2\tfrac{1}{2}n\log_2 n \tag{22}$$

(c) Solve harmonic equations - n tridiagonal equations, each of length $n2^{-\ell}$, are solved for the n harmonic amplitudes. The vectors now run horizontally and are of length n. The tridiagonal systems only involve variables from the last lines modified in stage (a) and Fourier transformed in stage (b). The time of execution is proportional to

$$t_c = 5(n_{\frac{1}{2}} + n)n2^{-\ell} \tag{23}$$

the coefficient five is appropriate for solution by Gauss elimination, taking into account that the immediate sub- and super-diagonals of the tridiagonal matrices are unity, and that the main diagonal is a constant.

(d) Fourier synthesis - on the same lines as stage (b) gives the solution on these lines. The FFT is used in the same way as in stage (b) giving

$$t_d = (n_{\frac{1}{2}} + n2^{-\ell})2\tfrac{1}{2}n\log_2 n \tag{24}$$

(e) Filling in - having found the solution on every 2^ℓ line in stage (d), fill-in takes place recursively. Each level, r, requires the formation of a right-hand side (2 operations) and the successive solution of 2^{r-1} tridiagonal systems. Vectors run vertically as in stage (a), and the time is proportional to

$$t_e = \sum_{r=1}^{\ell} (n_{\frac{1}{2}} + n2^{-r})(5 \times 2^{r-1} + 2)n \tag{25}$$

Evaluating the sums in Eqns. (21) to (25) we find the total time of execution per mesh point to be proportional to

$$n^{-2}t_{SERIFACR} = s + \left(\frac{n_{\frac{1}{2}}}{n}\right)q' \tag{26}$$

where

$$s = 4\ell + 4 + (1 + 5\log_2 n)2^{-\ell}$$

$$q' = 4\ell - 8 + 8 \times 2^\ell + 5 \times 2^{-\ell} + 5\log_2 n$$

The equal performance line between the algorithm with ℓ levels of reduction and that with $\ell + 1$ is easily found to be given by

$$\left(\frac{n_{\frac{1}{2}}}{n}\right) = \frac{(1 + 5\log_2 n)2^{-(\ell+1)} - 4}{4 + 8 \times 2^\ell - 5 \times 2^{-(\ell+1)}} \tag{27}$$

The form of Eqn. (27) suggests that a suitable parameter plane for the analysis of SERIFACR is the $(n_{\frac{1}{2}}/n, n)$ phase plane, and this is shown in Fig. 5. The equal performance lines given by Eqn. (27) divide the plane into regions in which $\ell = 0,1,2,3$ are the optimum choices. Lines of constant value of $n_{\frac{1}{2}}$ in this plane lie at 45 degrees to the axes, and the lines for $n_{\frac{1}{2}} = 20,100,2048$ are shown dotted in Fig. 5. These lines are considered typical for the behaviour, respectively, of the CRAY-1, CYBER 205, and the average performance of the ICL DAP. For practical mesh sizes (say $n<500$) we would expect to use $\ell=1$ or 2 on the CRAY-1, $\ell=0$ or 1 on the CYBER 205, and $\ell=0$ on the ICL DAP. The lower of the two values for ℓ applies to problems with $n<100$.

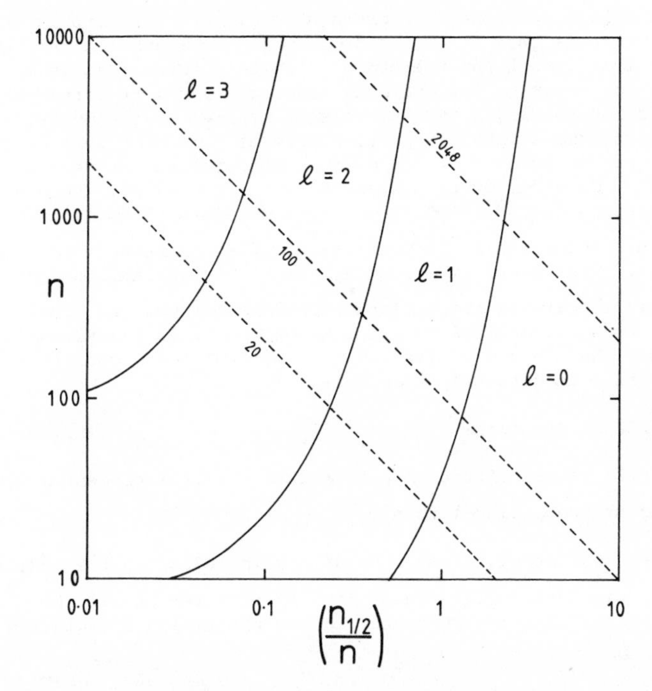

Fig. 5 The $(n_{\frac{1}{2}}/n, n)$ parameter plane for the SERIFACR(ℓ) algorithm. The solid lines delineate regions where the stated values of ℓ lead to the minimum execution time. The dotted lines are lines of constant $n_{\frac{1}{2}}$ corresponding to the CRAY-1 (=20), CYBER 205 (=100) and the average performance of the ICL DAP (=2048).

Temperton (1977) has timed a SERIFACR(ℓ) program on the CRAY-1 and measured the optimum value of ℓ=1 for n=32, 64 and 128. This agrees with our figure except for ℓ=128 where Fig. 5 predicts ℓ=2 as optimal. This discrepancy is probably because Temperton uses the Buneman form of cyclic reduction (see Hockney (1970)) which increases the computational cost of cyclic reduction and tends to move the optimum value of ℓ to smaller values.

For a given problem size (value of n) Fig. 5 shows more serial computers (smaller $n_{\frac{1}{2}}$) to the left and more parallel computers (larger $n_{\frac{1}{2}}$) to the right. We see therefore that the more parallel the computer, the smaller is the optimum value of ℓ.

In the SERIFACR algorithm the vectors are laid out along one or other side of the mesh and never exceed a vector length of n. It is an algorithm suited to computers that perform well on such vectors, i.e. those that have $n_{\frac{1}{2}} < n$, and/or which have a natural parallelism (or vector length) which matches n. The latter statement refers to the fact that some computers (e.g. CRAY-1) have vector registers capable of holding vectors of a certain length (64 elements in the CRAY-1). There is then an advantage in using an algorithm that has vectors of this length and therefore fits the hardware design of the computer. For example, the SERIFACR algorithm would be particularly well suited for solving a 64 × 64 Poisson problem on the CRAY-1 using vectors of maximum length 64; particularly as this machine is working at better than 80 percent of its maximum performance for vectors of this length. On other computers, such as the CYBER 205, there are no vector registers and $n_{\frac{1}{2}} \approx 100$. For these machines it is desirable to increase the vector length as much as possible, perferably to thousands of elements. This means implementing the FACR algorithm in such a way that the parallelism is proportional to n^2 rather than n. That is to say the vectors are matched to the size of the whole two-dimensional mesh, rather than to one of its sides. The PARAFACR algorithm that we now describe is designed to do this.

B. The PARAFACR Algorithm

Each of the stages of the FACR algorithm can be implemented with vector lengths proportional to n^2.

(a) Modify RHS - at each level, r, of cyclic reduction the modification of the right-hand side can be done in parallel on all the $n^2 2^{-r}$ mesh points that are involved. Hence the timing formula becomes

$$t_a = \sum_{r=1}^{\ell} (n_{\frac{1}{2}} + n^2 2^{-r})(3 \times 2^{r-1} + 2) \qquad (28)$$

(b) Fourier analysis - The $n2^{-\ell}$ transforms of length n are performed in parallel as in SERIFACR, but now we use a parallel algorithm, PARAFT, for performing the FFT with a vector length of n. The vector length for all lines becomes $n^2 2^{-\ell}$ and the timing equation is

$$t_b = (n_{\frac{1}{2}} + n^2 2^{-\ell}) 4\log_2 n \qquad (29)$$

The factor 4 replaces the 2½ in Eqn. (22) because extra operations are introduced in order to keep the vector length as high as possible in the PARAFT algorithm (see Hockney and Jesshope (1981), page 315). We also note that the factor n has moved inside the parentheses in comparing Eqn. (22) with (29), because the vector length has increased from $n2^{-\ell}$ to $n^2 2^{-\ell}$.

(c) Solve harmonic equations - the harmonic equations are solved in parallel as in SERIFACR, but we use a parallel form of scalar cyclic reduction, PARACR, instead of Gauss elimination for the solution of the tridiagonal systems (see Hockney and Jesshope (1981), page 289). For the special case of the coefficients previously noted, there are 3 parallel operations at each of $\log_2 n$

levels of scalar cyclic reduction. The vector length is $n^2 2^{-\ell}$ giving

$$t_c = (n_{\frac{1}{2}} + n^2 2^{-\ell}) 3\log_2 n \tag{30}$$

(d) <u>Fourier synthesis</u> - as stage (b)

$$t_d = (n_{\frac{1}{2}} + n^2 2^{-\ell}) 4\log_2 n \tag{31}$$

(e) <u>Filling in</u> - at each level, r, $n2^{-r}$ tridiagonal systems of length n are to be solved. Using PARACR as in stage (c) the vector length is $n^2 2^{-r}$. Afterwards a further two operations are required per point which may also be done in parallel giving

$$t_e = \sum_{r=1}^{\ell} (n_{\frac{1}{2}} + n^2 2^{-r})(3 \times 2^{r-1}\log_2 n + 2) \tag{32}$$

The time per mesh point for the PARAFACR algorithm is therefore proportional to

$$n^{-2} t_{PARAFACR} = s + \left(\frac{n_{\frac{1}{2}}}{n^2}\right) q'' \tag{33}$$

where

$$s = \tfrac{1}{2}(3\log_2 n + 1)\ell + 4 + (11 \log_2 n - 4)2^{-\ell}$$

$$q'' = 4\ell + (3\log_2 n + 1)(2^{\ell} - 1) + 11 \log_2 n$$

The equal performance line between the level ℓ and ℓ + 1 algorithms is given by

$$\left(\frac{n_{\frac{1}{2}}}{n^2}\right) = \frac{(11 \log_2 n - 4)2^{-(\ell+1)} - \tfrac{1}{2}(3\log_2 n + 1)}{4 + (3\log_2 n + 1)2^{\ell}} \tag{34}$$

The form of Eqn. (34) leads us to choose to plot the results for the PARAFACR algorithm on the $(n_{\frac{1}{2}}/n^2, n)$ parameter plane, and this is done in Fig. 6. We find that the equal performance lines are approximately vertical in this plane, and conclude that $\ell = 2$ is optimal for $n_{\frac{1}{2}} < 0.1n^2$, $\ell = 1$ for $0.1n^2 < n_{\frac{1}{2}} < n^2$ and, $\ell = 0$ for $n_{\frac{1}{2}} > n^2$. There are no circumstances when more than two levels of reduction are worth while, thus justifying our use of the unstabilised FACR algorithm (see Hockney and Jesshope (1981), page 348). In particular, for a processor array with as many or more processors than mesh points ($N \geqq n^2$), we take $n_{\frac{1}{2}} = \infty$ and find $\ell = 0$. This case corresponds to the solution of a 64 × 64 problem on the ICL DAP which is an array of 64 × 64 processors. The dotted line for $n_{\frac{1}{2}} = 100$ is shown in Fig. 6. corresponding to the CYBER 205. For all but the smallest meshes (i.e. for $n \geqq 30$) we find $\ell = 2$ optimal. The line for $n_{\frac{1}{2}} = 20$ is also given, from which we conclude that $\ell = 2$ is optimal in all circumstances if this algorithm is used on the CRAY-1.

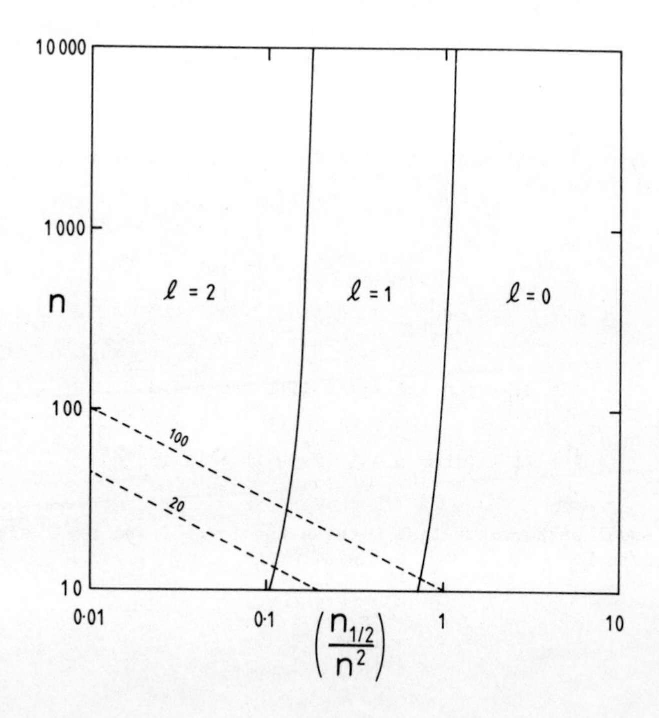

Fig. 6 The $(n_{\frac{1}{2}}/n^2, n)$ parameter plane for the PARAFACR(ℓ) algorithm. Notation as in Fig. 5.

C. SERIFACR/PARAFACR Comparison

So far we have considered the choice of the best value of ℓ for each algorithm. Having optimised each algorithm we now consider which is the best algorithm to use. This is done by plotting $t_{SERIFACR}$ and $t_{PARAFACR}$ against $(n_{\frac{1}{2}}/n)$ for a series of values of n, in order to determine approximately which algorithms abut each other in different parts of the parameter plane. One can then calculate the equal performance line between PARAFACR(ℓ) and SERIFACR(ℓ') from

$$\left(\frac{n_{\frac{1}{2}}}{n}\right) = \frac{a - b}{c - d} \tag{35}$$

where

$$a = \tfrac{1}{2}(3\log_2 n + 1)\ell + 4 + (11 \log_2 n - 4)2^{-\ell}$$

$$b = 4\ell' + 4 + (1 + 5\log_2 n)2^{-\ell'}$$

$$c = 4\ell' - 8 + 8 \times 2^{\ell'} + 5 \times 2^{-\ell'} + 5\log_2 n$$

$$d = \left(\frac{4\ell + (3\log_2 n + 1)(2^{\ell} - 1) + 11 \log_2 n}{n}\right)$$

The interaction of the two algorithms is shown in Fig. 7 on the $(n_{\frac{1}{2}}/n, n)$ parameter plane. This division between the two algorithms is about vertical in this plane showing that SERIFACR is the best algorithm for smaller $n_{\frac{1}{2}} < 0.4n$ (the more serial computers), and the PARAFACR is the best for larger $n_{\frac{1}{2}} > 0.4n$ (the more parallel computers). Lines of constant $n_{\frac{1}{2}}$ are shown for the CRAY-1 and CYBER 205. We conclude that SERIFACR should be used on the CRAY-1 except for small meshes with $n < 64$ when PARAFACR(2) is likely to be better. On the CYBER 205, PARAFACR is preferred except for very large meshes when SERIFACR(2) $(300 < n < 1500)$ or SERIFACR(1) $(n > 1500)$ is better.

5. CONCLUSIONS

The optimum choice of algorithm for the solution of Poisson's equation on a parallel computer is found to depend on the ratio of the parallelism of the computer (as measured by its half-performance length) to the size of the finite difference mesh. Two implementations of the FACR(ℓ) algorithm have been considered, and for each the optimum value of ℓ is predicted as a function of parallelism and mesh size. The results are presented in a useful graphical form by drawing regions in a convenient parameter plane where each value of ℓ is optimal. In both cases we conclude that less cyclic reduction (lower ℓ) should be performed the more parallel is the computer. Finally, the results for the two implementations are combined to show the conditions in which each should be used. Not unnaturally, we find that the implementation with

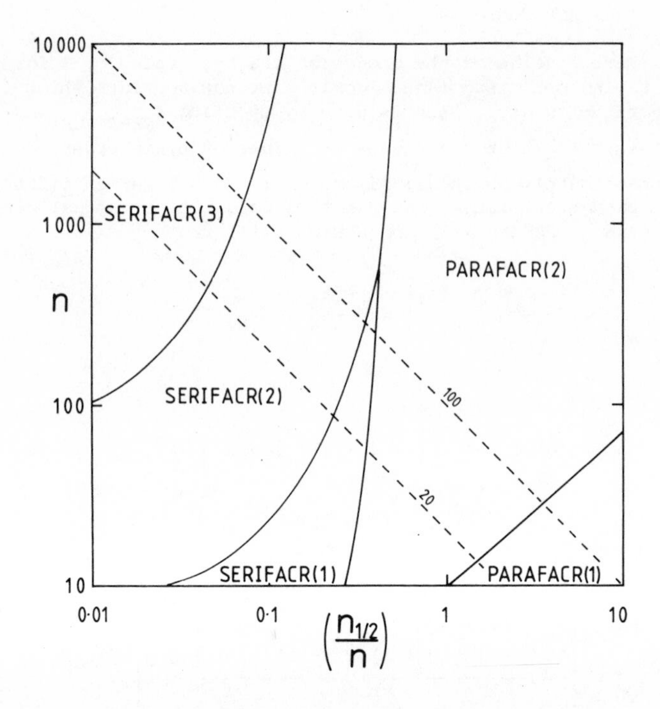

Fig. 7 Comparison between the SERIFACR(ℓ) and PARAFACR(ℓ), showing the regions of the $(n_{\frac{1}{2}}/n, n)$ parameter plane where each has the minimum execution time.

the smallest vector length (or algorithmic parallelism) SERIFACR, is most suitable for computers with low hardware parallelism (i.e. the more serial with low $n_{\frac{1}{2}}$), and that the implementation with the longest vector length PARAFACR, is most suitable for computers with high hardware parallelism. The final figure gives quantitative predictions for the conditions in which each implementation should be used.

All the above conclusions are based on simplifying assumptions discussed in the introduction, and can only be taken as indicative of the results that may actually be obtained with real programs on real computers. In practice the best program will depend on the relative efficiency of the code used for different subroutines (e.g. the tridiagonal solver and the FFT routine), and the extent to which memory bank conflicts and routing delays have been minimised. Our analysis is ignorant of these influences. It addresses only the effect of vector length on the execution of an algorithm. Nevertheless this is one more effect than is usually considered.

The best practice is probably to write a program for each implementation with variable ℓ and determine empirically the optimum algorithm and level of reduction. Our graphs can be used as a guide to which algorithm and value of ℓ is likely to be the best in each region of parameter space.

ACKNOWLEDGEMENTS

The author wishes to thank Chris Jesshope, Jim Craigie and Edward Detyna for help in clarifying the ideas in this paper, and Knut Mörken for pointing out several misprints in the original manuscript.

REFERENCES

Hockney, R.W. and Jesshope, C.R. (1981) Parallel computers - Architecture, Programming and Algorithms. Adam Hilger, Bristol. (Distributed in North and South America by Heyden and Son Inc., Philadelphia.)

Calahan, D.A. and Ames, W.G. (1979) Vector Processors: Models and applications. *IEEE Trans. Circuits and Systems*. CAS-26, pp. 715-726.

Kuck, D.J. (1978) Computers and Computations. Wiley, New York.

Potter, D. (1973) Computational Physics. Wiley, London.

Hockney, R.W. and Eastwood, J.W. (1981) Computer Simulation Using Particles. McGraw-Hill, New York.

Hockney, R.W. (1965) A fast direct solution of Poisson's equation using Fourier analysis. *J. Assoc. Comput. Mach.*, **12**, pp. 95-113.

Hockney, R.W. (1970) "The potential calculation and some applications". *Methods Comput. Phys.*, **9**, pp. 135-211.

Swarztrauber, P.N. (1978) The method of cyclic reduction, Fourier analysis and the FACR algorithm for the discrete solution of Poisson's equation on a rectangle. *SIAM Rev.*, **19**, pp. 490-501.

Temperton, C. (1977) Fast Fourier transforms on the CRAY-1. Infotech State of the Art Report:Supercomputers (C.R. Jesshope and R.W. Hockney, Eds.), **2**, pp. 359-379.

PROGRAM DEVELOPMENT SOFTWARE FOR ARRAY PROCESSORS

J.M. Marsh

(System Software Factors, Reading)

ABSTRACT

A wide range of software is provided by different Array Processor manufacturers to assist the user in developing applications. This paper is the result of a survey of the program development software provided for various computers. The different approaches to the problem of developing programs will be discussed under a number of software categories, and several specific examples will be given. The paper concludes with an appendix giving a brief overview of the array processors used in the examples.

1. INTRODUCTION

The last ten years has seen the development and manufacture of a number of computers and attached processors which can be classified as Array Processors. These machines come in a great variety of prices and sizes, and encompass a wide range of hardware architectures and technologies. The common goal across this range is to deliver extra computational power to the user, and this involves providing program development software.

An interesting characteristic of Array Processor architectures is that, at the lowest level, they are all very difficult to program. Invariably some or all of the extra computational power is achieved by architectural techniques, such as pipelining and parallelism. As a result programming at the machine level is very difficult in comparison with Assembler coding for a typical minicomputer. This characteristic puts added importance on the provision of a high level language or environment for the user. Unfortunately, if it is difficult to program the Array Processor in Assembler, then it is difficult to implement an efficient high level language for the Array Processor.

The result is that the program development software available for such machines varies enormously in its quality and in the level of support provided. Not surprisingly, the more expensive the computer is, the more software is generally available. The exact type and effectiveness, however, is of paramount importance to a potential user, since the successful implementation of his application depends upon the software provided.

2. PROGRAM DEVELOPMENT SOFTWARE CATEGORIES

Despite the many and varied approaches that Array Processor manufacturers have taken to the problem of providing program development software, it is possible to generalise, and discuss the software under a number of broad categories. For the purpose of illustration by example, several Array Processor manufacturers will be included specifically. Appendix A contains a list of the Array Processors concerned and a brief description of the hardware architectures involved.

2.1 *Application Libraries*

Most, if not all, of the Array Processor manufacturers offer a library of highly optimised assembler coded routines to perform a range of vector and matrix operations. Usually, these subroutines are supplemented by Fast Fourier Transform routines, as well as specialist algorithms for signal processing, image processing, simulation applications and others.

Use of these libraries varies. At the lower end of the market, the library may provide the building blocks from which to implement the computational part of the application. In such cases, the final product has the form of a host program making a number of subroutine calls to library routines which execute in the Array Processor. Alternatively, when an optimising Fortran compiler is available for the Array Processor, library calls may only be used to invoke substantial routines, such as matrix inversion or an equation solver.

2.2 *"Chainers"*

When a Fortran compiler is not available for the Array Processor, an alternative is needed for repeated library calls from the host computer, as the overheads involved in alternating control between the two processors may be prohibitive. A common approach is to provide a technique that allows several library calls to be "chained" together, and invoked as a unit in the Array Processor. The details of how this is accomplished vary. A popular scheme, used by CSPI and ANALOGIC, is to construct at run-time a number of control or descriptor blocks describing the routine to be called and its arguments. These blocks can then be chained together, and passed to the AP, which executes all the routines specified by the chain before returning control to the host. An alternative approach is to provide a translator which will generate Assembler code from a file containing a number of library subroutine calls. The resulting Assembler module can then be called as a unit from the host. An example of this is the FPS Vector Function Chainer, which can also support other simple operations, such as integer arithmetic and control statements. They are completely superseded by high level languages.

2.3 *Assemblers*

Most manufacturers provide an Assembler language. However, programming an Array Processor at the machine level is usually difficult, tedious and fraught with dangers. Efficient use of architectural features usually requires careful management of registers and exact timing and coordination of interacting functional units. A further problem is that considerable effort may go into producing a highly optimised Assembler routine for a particular task which may be difficult to document well, and may prove impossible to modify, requiring a complete rewrite should the specification of the task change slightly. Particular difficulties also arise if the parallel processors operate asynchronously.

The form of the Assembly language depends greatly on the architecture. Some Array Processors, like the FPS AP120B, execute a microcoded instruction, where a long instruction word has a number of fields each controlling a discrete processor. Such machines usually operate synchronously, with each processor completing its allocated task by the end of a clock tick. Asynchronous architectures, like the CSPI MAP, may require that each processor be programmed independently, and some system of "handshakes" or semaphores is needed for synchronisation. The Assembler for the ANALOGIC AP400 resembles that of a conventional minicomputer, and the vector hardware is not directly programmable. Instead, a number of instruction sequences for the vector hardware are preprogrammed into PROM, and can be invoked using macro calls at relevant stages in the Assembler code.

2.4 High Level Languages

Fortran compilers are provided for several Array Processors, particularly for the more expensive "supercomputers". Indeed, the provision of a high-quality optimising Fortran compiler is central to the program development strategy for machines such as the CRAY. In general, it appears that expensive machines being sold into end-user environments require good Fortran support. At the other end of the scale, the inexpensive Array Processors aimed at the volume OEM† market need only the minimum of program development tools, since an OEM customer will develop an application once, and then sell several copies of the resulting hardware and software system.

Other high level languages are available; for example, a PASCAL compiler has been announced for the CRAY. Another approach is to provide a high level environment. The Applied Dyanmics AD10 is programmed using a simulation environment whereby simulation applications are built up from component parts using a simulation language.

The CRAY Fortran compiler illustrates the problems that need to be overcome. It supports ANSI '66 and '77 Fortran language standards, as well as being "dialect tolerant" to assist users porting Fortran applications from a range of other machines. Its optimisation strategy consists mainly of recognising a range of DO loops and other constructs as being "vectorisable" and generating code to use the vector hardware accordingly. Constructs that are not vectorisable execute in the scalar hardware at a slower rate. Certain classes of DO loop are not vectorisable, such as those containing complex IF statements, although improvements are continually being made. For example:

```
        DO   10   I  =  1,100
        SUM  = SUM + A(I) * B(I)
10      CONTINUE
```

was not vectorised a year ago, because the result, SUM, was a scalar rather than a vector. The latest CRAY Fortran compiler will recognise this situation, and generate code to perform a vector multiply operation resulting in a temporary vector. This vector is then segmented into a number of small vectors, of length 8, say. A series of vector add operations results in a single short vector, and the result SUM is then computed in the scalar hardware. In addition to the automatic vectorisation of a range of constructs, the compiler reports successful and unsuccessful attempts at vectorisation, so that the user can rewrite sections for extra speed.

† Other Equipment Manufacturer

Other Fortran compilers offer a different approach. The DAP Fortran compiler supports language extensions which facilitate the generation of optimised code. These language extensions precisely reflect the DAP architecture, encouraging the user to program applications for the DAP most efficiently. The price to pay, here, is loss of portability and increased programming effort. The main benefit is that applications programmed in DAP Fortran are typically as efficient as they would be programmed in DAP Assembler.

Floating Point Systems provide Fortran compilers for both the FPS164 and the AP120B. The latter compiler supports a computational subset of Fortran '66, with all I/O statements excluded. Using the AP120B at the Fortran level involves splitting the application program into two halves, a computational portion for the AP120B and a driving portion that resides in the host. Each program portion is built using the relevant compiler and linker according to the destination machine. Communication between the two portions is achieved using data transfer utilities, the use of which is mainly invisible to the user. The compiler for the FPS164 supports the Fortran '77 language level. It can be used in two ways. The first is entirely comparable to the AP120B compiler, with I/O restricted to the host and mainly automatic communication between the two processors. More recently, FPS have implemented an environment whereby complete Fortran jobs can run in the FPS164 with I/O being directed to the local attached disks.

2.5 Ancillary Software

As well as software involved directly with the development of programs, the support software provided should also be considered. Such software includes hardware simulators, debugging aids and executives for the Array Processor.

All manufacturers supply some software to control the Array Processor. This may be only a set of routines to start and stop the Array Processor functions under the control of a host program, or may be a free standing executive for the Array Processor, scheduling jobs for execution within the Array Processor. Floating Point Systems provide examples of both types. The interface software for the AP120B, named APEX (Array Processor Executive), consists of a number of routines to control data transfer between the two processors, to interrogate the interface and hardware status and to control the running of the AP120B. Contrastingly, the FPS164 may be run under the control of the SJE (Single Job Executive) which supports a simple file structure on the FPS164 local disks, and stages data transfer between the FPS164 and the host periphals.

Support for debugging programs may also be an important consideration. Debugging tools are usually available, but are frequently crude and only give access to the low-level functions of the hardware - registers, memory locations and so on. High level debuggers are available, from CRAY and FPS for example, and these allow Fortran programs to be tested in the Array Processor under the control of the user. Using such tools, the symbols and structures of the Fortran program are available, rather than low-level hardware functions.

3. CONCLUSION

The main conclusion to be drawn from this survey is that the program development software provided for array processor users is generally quite primitive compared to that typically available for a modern mini-computer or mainframe. The range and type of software available depends entirely upon the market that the Array Processor is aimed at. This is illustrated by considering Floating Point Systems software, and how it has evolved over the last ten years. Initially, the AP120B was aimed at the quantity OEM market, and the software was limited to Application libraries and Assembler. Subsequently, FPS have achieved some success in selling the AP120B to more end users, and have had to develop Fortran language support to satisfy this market. Most recently FPS has moved "up market", with the FPS164 attached processor, and the software available includes high level debugging tools, Fortran 77 compiler and the Single Job Executive.

APPENDIX A

This appendix contains a list of the Array Processor manufacturers mentioned in this paper, together with a brief overview of the associated hardware architecture and software.

A.1	ANALOGIC	16 bit fixed point vector hardware ("pipes") plus conventional scalar hardware. Pipes activated from scalar hardware by macro calls. Pipes programmed in PROM. Assembler and Application library provided with chaining set up at run time.
A.2	APPLIED DYNAMICS	multiple discrete processors controlled by micro-code. 16 bit fixed point arithmetic. Application library, Assembler and high level simulation environment supplied.
A.3	CRAY RESEARCH	multiple pipelined functional units in Vector hardware performs 64 bit floating point arithmetic on multiple banks of registers. Scalar hardware for addressing and non vectorisable operations. Single instruction issued every clock cycle, but vector function will perform up to 64 operations, once instruction issued, in parallel with other functional units. Software includes Fortran 66 and 77 compilers, Pascal, high level debugger, batch executive, Assembler, application libraries, monitoring utilities.
A.4	I.C.L.	Distributed Array Processor in a 64 x 64 lattice of bit-processors, each with 4096 bits of memory. Single instructions are issued to be performed by many of the 4096 processors in parallel. Software includes Assembler, application libraries and specially extended DAP-Fortran compiler.
A.5	F.P.S.	In general, multiple pipelined functional units controlled by microcoded instructions, so several sub-instructions issued in parallel. AP120B and FPS-100 are 38 bit floating point, while FPS164

is 64 bit floating point arithmetic. Software includes Application libraries, Assembler, Fortran compilers and single Job Executive for the FPS164.

A.6 C.S.P.I. Multiple functional units programmed independently running, in general, asynchronously. 32 bit floating point arithmetic. Assembler, Application libraries and a chaining system.

ON MAKING THE NAG RUN FASTER

D.H. McGlynn and L.E. Scales

(Department of Computer Science, University of Liverpool)

ABSTRACT

This paper concerns user experience in improving the performance of NAG library FORTRAN routines for the CRAY-1S. Following a brief discussion of CRAY hardware and software features which need to be borne in mind, some preliminary results from the linear algebra section of the NAG library will be presented. The relative advantages of using the BLAS (Basic Linear Algebra Subroutines) in CRAY SCILIB and of using modified in-line code will be discussed, and some timings given for the various approaches. Some of the limitations imposed by the need for the new version of the library to be as consistent as possible with the old will be indicated. Finally, a number of case studies will be presented.

1. INTRODUCTION

The results presented herein derive from work undertaken with the aid of a CASE studentship awarded by SERC, the industrial co-sponsor being CRAY Research (UK) Ltd. The aim of the work is to generate a version of the CRAY implementation of the NAG library in which suitable routines have been improved with regard to the performance of vector operations. The examples of code are copyright of Numerical Algorithms Group Ltd. The authors wish to express their thanks to all these bodies for their assistance, and especially to J.J. Du Croz of NAG Central Office, who was a constant source of help, advice and encouragement.

The techniques used are not novel, and are quite well documented (CFT manual part 3 section 2; Higbie (1979); Sydow (1979)). There is no attempt to implement radically new algorithms, but rather to explore the extent of improvements possible within the framework of existing codes.

Timings were derived using the CFT function SECOND, and timings obtained on different occasions varied little. The timing runs were made on the CRAY at the Daresbury Nuclear Physics Laboratory, which at that time had a memory of half a million words, with the CFT 1.09 compiler. Other software such as SCILIB, COS and CAL were initially release 1.09, but 1.10 was implemented during more recent work. There were some changes to CFT 1.09 during the period when these results were generated, but as far as can be seen, they were not such as to significantly affect the results.

2. RELEVANT HARDWARE FEATURES OF THE CRAY-1S

2.1 Vector Registers, Instructions and Chaining

The CRAY has eight vector registers, each of 64 words. Most instructions have a vector version, so that a single machine instruction may initiate up to 64 operations. There is also a set of functional units for vector processing; some of these are specific to vector processes, but the arithmetic functional units (add, multiply and reciprocal approximation) are shared by the V (vector) and S (scalar) registers.

The functional units are pipelined and, after start-up, results issue at the rate of one per machine cycle, which is 80 million per second. As the functional units may operate independently in parallel, one may have several vector processes in progress at the same time, each producing results at this rate, so that megaflop rates in excess of 80 are attainable.

When several vector instructions are issued in succession, there is the possibility of chaining, in which the results from one vector operation are fed directly to the functional unit which performs the next vector operation, without having to wait for the preceding process to be completed. In order for chaining to occur, it is necessary for the second vector instruction to be issued before the so-called chain slot of the first, and for any other operands to be available in vector registers. If conditions are right, then chaining is effected automatically by the hardware.

The CRAY FORTRAN compiler (CFT) analyses Fortran code, converts innermost DO-loops to vector operations where possible, and compiles vector instructions to implement them. However, the analysis is not sufficiently powerful to make the best use of the vector-processing features in all circumstances and it is often possible to improve substantially on the compiled object code by resorting to CRAY Assembly Language (CAL).

2.2 Memory access conflicts

The CRAY memory is organised in 8 or 16 banks (8 in Serial 28, the machine currently at Daresbury) in such a way that contiguous addresses are in successive banks. Once a bank is accessed, it may not be accessed again for four clock periods. Therefore, though there is no direct overhead in accessing memory with strides other than unity, the use of multiples of four or eight (or on a 16-bank machine 8 or 16) gives rise to memory access conflicts. These slow down the reading from memory, and furthermore prevent the access chaining with any succeeding vector process. Thus wherever possible one should access matrices by column (hence contiguous address locations) rather than by row, especially when one has no prior knowledge of the dimensions of the matrices to be operated upon.

2.3 Instruction issue

The CRAY has four instruction buffers into which object code is read from memory before being executed. Each holds 16 words of 64 bits, and an instruction may be one or two parcels of 16 bits, so that each buffer may hold up to 64 instructions. All is well until a branch instruction is met. This is a two-parcel instruction, the first identifying it as a branch, and the second giving the destination. It normally takes 5 clock periods to execute, but depends upon the position in the buffers of the second parcel of the branch instruction and of the instruction to be executed after branching. The maximum time to execute the branch is 25 clock periods. When working in FORTRAN, this is largely beyond our control, but it does point to the necessity of good programming practice with the avoidance of too much jumping about, and may occasionally explain some anomalous results.

3. SOFTWARE CONSIDERATIONS

3.1 Double precision

There are no hardware double precision facilities on the CRAY, and all double precision is performed in software. Though there are vector versions of the appropriate routines, they do involve very high overheads. One is therefore advised to restrict additional precision to those cases where it is absolutely essential.

3.2 Subroutine overheads

These are high on the CRAY. Therefore, trivial auxiliary routines are probably best converted into in-line code wherever possible. It was found for example that moving the scaling out of F03AFF into a more efficient auxiliary routine made little apparent improvement, and the same exercise in F03AHF caused a slowing down.

3.3 SCILIB routines

The BLAS (Basic Linear Algebra Subprograms) (Lawson et al. (1979)) have been coded in CAL (Petersen (1979)) and included in the standard SCILIB library on the CRAY, and their potential for use in the current work was explored.

A simple test was run, in which timings, at vector lengths increasing at intervals of 10 from 10 to 750, with a memory stride of unity to avoid memory access clashes, were extracted for those BLAS under consideration and for the corresponding in-line FORTRAN code. No attempt was made to optimise the usage of instruction buffers, but rather the test routines were so structured as to be poor in this respect, with a view to reflecting the conditions under which any substitution of code might be made. Some typical results are presented in the tables below; they differ somewhat from those given by Higbie, but his work was performed some time ago, and there have since been changes to the CRAY software; furthermore, no attempt was made to achieve ideal conditions for the testing of the BLAS.

Results of timing tests on selected BLAS (Times in microseconds):

(BLAS whose names begin with S operate on REAL data; those whose names begin with C operate on COMPLEX data.)

Dot products SDOT, CDOTC and CDOTU:

Vector length	10	20	30	40	50	60	70	80	90	100
SDOT	4.26	4.50	4.75	5.00	5.25	5.50	6.51	6.61	6.74	6.93
In-line code	5.05	5.68	6.30	6.93	7.55	8.18	9.26	9.80	10.42	11.05

Vector length	10	20	30	40	50	60	70	80	90	100
CDOTU	5.66	6.50	7.38	8.25	9.13	10.00	11.90	12.64	13.51	14.39
In-line code	8.33	15.95	23.58	31.20	38.82	46.45	54.07	61.70	69.32	76.95

Vector length	10	20	30	40	50	60	70	80	90	100
CDOTC	5.41	6.25	7.13	8.00	8.88	9.75	11.65	12.39	13.26	14.14
In-line code	8.50	16.37	24.25	32.12	40.00	47.87	55.75	63.62	71.50	79.38

SAXPY, CAXPY (y = ax+y; a scalar, x and y vectors.):

Vector length	50	100	150	200	250	300	350	400	450	500
SAXPY	4.94	7.25	9.56	11.06	13.75	16.06	18.38	20.69	23.21	24.88
In-line code	3.60	6.43	9.25	12.14	14.58	17.40	20.23	23.05	26.16	28.38

Vector length	20	40	60	80	100	120	140	160	180	200
CAXPY	5.05	6.80	8.55	11.13	12.88	14.63	17.24	18.95	20.70	23.38
In-line code	3.91	5.91	7.91	10.96	12.91	14.91	18.01	19.91	21.91	25.06

SCOPY, CCOPY (y = x; x and y vectors):

Vector length	100	200	300	400	500	600	700
SCOPY	4.78	7.68	10.25	13.10	15.77	18.62	21.30
In-line code	3.53	6.34	8.95	11.80	14.47	17.32	20.00

Vector length	100	200	300	400	500	600	700
CCOPY	7.83	13.37	18.50	23.95	29.17	34.62	39.85
In-line code	6.30	11.90	16.97	22.47	27.65	33.10	38.32

It would appear from these results that SASUM, SCASUM, SDOT, CDOTC, CDOTU, SNRM2, and SCNRM2 out-perform in-line FORTRAN even at small vector lengths, and their advantage increases with vector length. On the other hand, CAXPY, SCOPY, CCOPY, CSWAP and SSCAL never recoup the overhead of the subroutine call. SAXPY, SSWAP, CSSCAL and CSCAL each have a vector length above which they are superior to in-line code, but this is generally quite large, and it seems unlikely that the general user will benefit should we substitute them into NAG routines. Thus, only the BLAS in the first group have been used in modifications adopted so far.

4. GUIDELINES

The aim of the project has been to produce code which ideally makes the best possible use of the vector registers and functional units of the CRAY; this includes chaining of vector operations whenever possible.

However, this aim has to be constrained by a number of considerations, since the code in question is contained in a widely-used subroutine library.

It was agreed that the user-interface to the routines should not be changed, so that the specification of the routines, as documented in the NAG Library Manual, would still be applicable.

It was also agreed that the modifications to the code would conform to standard Fortran, at least in the first phase of the work, in order that the code could eventually be transported to other vector-processing machines. Thus machine-specific extensions to Fortran, such as the conditional vector merge functions, have been avoided. No CAL coding has been included yet, although this may be undertaken later in order to come closer to the optimal performance of the machine. However calls to the BLAS were permitted and also the use of the CRAY compiler directive CDIR$ IVDEP.

Within the NAG routines the code could be re-organised quite freely; in some cases calls to auxiliary routines have been removed (especially to the inner product routines X03AAF and X03ABF), while in other places new auxiliary routines have been introduced.

5. MODUS OPERANDI

Having decided that a routine was a suitable case for treatment, the necessary editing was done using UPDATE. Thereupon, the NAG example program was run in order to check that the routine behaved more or less as it should, and this was followed by the NAG stringent test program. Any differences in results were then carefully weighed. FLOWTRACE was used to give some indication of the benefits derived from the changes. This was but a rough guide to the improvement, as the test matrices are of low order, but even so in good cases reductions of a factor of 5, 6 or more in the time spent within the routine were observed. When a modification appeared to be performing well numerically, a simple timing program was written and run to derive speed-up figures such as those quoted below.

6. CANDIDATES FOR CHANGE

In the codes studied so far, the major areas for improvement have been sections of code in which inner loops do not vectorise, for example the search for a pivot:

```
      DO 20 I=1,N
         IF (X.GE.ABS(A(I,J))) GO TO 20
         X = ABS(A(I,J))
   20 CONTINUE
```

or the scaling of a determinant:

```
  100 IF (D1.GT.1.0) GO TO 120
      D1 = D1*0.0625
      ID = ID+4
      GO TO 100
  120 .....
```

and codes which only partly vectorise, as in the calculation of inner products:

```
      DO 20 I=1,N
   20 X = X + A(J,I)*B(I,J)
```

On the CRAY, the last of these is dealt with in a rather individualistic manner. The use of a recursive vector add instruction (i.e., the results are sent to one of the operand registers) causes the 64 elements in that register to be reduced to 8 partial sums as a vector process; these sums must then be further accumulated in the scalar registers, so that the calculation of inner products may be considered to be a partial vector process on the CRAY. This type of reduction works only for types REAL and COMPLEX, not for DOUBLE PRECISION.

As shown earlier, the use of CAL-coded BLAS gives some improvement over in-line dot-products, but there is a further approach which one may take when dealing with inner products on successive rows/columns of matrices, as for example when solving sets of linear equations. In this, one rearranges the arithmetic operations in such a way as to allow the accumulations to be done in parallel. When this is done, all operations are effected entirely as vector processes, as there is no vector reduction involved. Thus for example the code:

```
      DO 180 K=1,N
         DO 40 I=K,N
            Y = 0.0
            DO 20 J=1,K-1
               Y = Y + A(I,J)*B(J,K)
   20       CONTINUE
            A(I,K) = A(I,K)-Y
   40    CONTINUE
  180 CONTINUE
```

may be rewritten:

```
      DO 180 K=1,N
         DO 40 J=1,K-1
            DO 20 I=K,N
               A(I,K) = A(I,K) - A(I,J)*B(J,K)
   20       CONTINUE
   40    CONTINUE
  180 CONTINUE
```

(the operation in the innermost loop is referred to as a SAXPY type operation, after the BLAS of that name.)

Such a rearrangement gives code which is faster than that using the SCILIB routines, and often has the further advantage of accessing matrices within the inner loop entirely by column, not by row, so that one is retrieving data from contiguous memory addresses, hence avoiding memory access conflicts. Typical results, when performing operations on a triangular system, as in the above codes, may be seen below:

Order of matrix, N:	10	60	100
Time using CFT inner products:	53.94	443.8	934.8
Time using SDOT:	42.88	355.3	737.0
Time using SAXPY type code:	21.00	206.3	487.2

Figures for the complex case show an even greater improvement:

Order of matrix, N:	10	60	100
Time using CFT inner products:	51.35	1412.0	3829.0
Time using CDOTC:	57.36	742.0	1775.0
Time using SAXPY type code:	30.10	355.1	878.3

7. MODIFICATIONS TRIED

7.1 Abandoning Double Precision

Many of the NAG Linear Algebra routines perform the bulk of their computation via calls to an auxiliary inner product routine X03AAF (or X03ABF for complex inner products). This has a LOGICAL parameter SWITCH: if SWITCH is .TRUE., the inner product is computed in additional precision, otherwise in basic precision. In the single precision base version of the NAG Library the parameter SWITCH is set to .TRUE. in almost all calls to X03AAF. This assumes that the use of double precision would not imply a serious overhead, and would be worthwhile to achieve the improvement in numerical accuracy. However, on many machines the use of additional precision does involve a considerable overhead; this is nearly always true, for example, when the basic precision of computation in the library is double precision, and setting SWITCH to .TRUE. forces the use of quadruple precision. On such machines the parameter SWITCH has been changed to .FALSE. in all calls to X03AAF except where it is essential for the correct performance of the algorithm (e.g. in routines which perform iterative improvement of the solutions of linear equations). With the agreement of NAG, it was decided to adopt the same approach on the CRAY, given the high degree of precision available in single precision working, and the very high overheads of using double precision (see 3.1 above and 7.4 below).

7.2 Use of the BLAS

In the light of the timings given earlier, the effect of replacing X03AAF by SDOT, and X03ABF by CDOTU was tried, and the results were encouraging. The SAXPY-type solution was then tried, with even better results, and it was apparent that the BLAS were not likely to be as useful as was first thought.

7.3 Other possibilities

A variety of other changes were attempted, including the moving of the quite unvectorisable scaling loops (as illustrated earlier) into a subroutine where the same effect was achieved by masking and shifting; replacing the search loops (again as illustrated earlier) with the functions ISAMAX et al. from SCILIB; eliminating unnecessary operations by tidying up some of the indexing; eliminating unnecessary subroutine calls by shifting special cases such as first and last rows out of main loops; even to some extent completely restructuring the code while maintaining the same underlying algorithm. All of these had little more than marginal effect, and left one with the feeling that a few relatively straightforward changes gave much improvement, while to extract the last ounce of speed took far more time and usually gave results that were not commensurate with the effort.

7.4 A Few Results

The effectiveness of the various strategies can be judged from the results in the following tables. The first relates to FO3AFF, which performs Crout factorisation on a real matrix, the second to FO3AHF which does the same in the complex case. Speed up is given by the ratio

$$\frac{\text{Time taken by unmodified routine to factorwise order 99 matrix}}{\text{Time taken by modified routine on the same task}}$$

where the times are derived as explained earlier.

Similar changes were made to the corresponding equation solvers FO4AJF and FO4AKF, first trying the effect of using BLAS in place of XO3... routines, then using a SAXPY-type reconstruction of the code, and the results are given in the third and fourth of the following tables.

Code to illustrate the final changes to FO3AFF will be found in the appendix.

Results obtained for FO3AFF:

Changes implemented:	Speed up
Call XO3AAF with SWITCH set .FALSE.	7.53
Use SDOT in place of XO3AAF	38.50
Further modifications	40.03
Use SAXPY type oerations	53.95
Restructured SAXPY-type code	63.53
Further modifications	67.18

Results obtained for FO3AHF:

Changes implemented:	Speed up
Call XO3ABF with SWITCH set.FALSE.	17.50
Use CDOTC in place of XO3ABF	75.8
Further modifications	80.0
Use SAXPY type operations	100.6
Restructured SAXPY-type code	119.1

Results obtained for FO4AJF;

Changes implemented	Speed up
Use SDOT in place of XO3AAF	31.52
Use SDOT and make other minor changes	34.00
Use SAXPY-type operations	40.68
Use SAXPY-type with other changes	40.89

Results obtained for FO4AKF

Changes implemented	Speed up
Use CDOTU in place of XO3ABF	54.19
Use CDOTU with other changes	56.82
Use SAXPY-type operations	63.67
SAXPY-type with other changes	63.71

8. GENERAL RESULTS AND CONCLUSIONS

The work described here has concentrated on the FO3 and FO4 chapters of the NAG Library which are concerned with the solution of systems of linear equations and the associated matrix factorizations. The following figures give the best results achieved so far in these chapters (in each case we give the speed up as defined earlier):

FO3AEF	Cholesky factorisation	63.12
FO3AFF	Crout factorisation (real)	67.18
FO3AGF	Banded Cholesky factorisation	2.56
FO3AHF	Crout factorisation (complex)	119.14
FO4AGF	Solution of equations after factorization by FO3AEF	40.10
FO4AJF	Solution of equations after factorization by FO3AFF	40.89
FO4AKF	Solution of equations after factorization by FO3AHF	63.71
FO4ALF	Solution of equations after factorization by FO3AGF	4.80
FO4ANF	Solution of equations after QR factorization	12.37
FO4AQF	Similar to FO4AGF but using packed storage	2.70
FO4ARF	Similar to FO4AJF but for single right hand side	66.15

Clearly, then, the range of improvements attained was wide. Furthermore, efforts to improve the iterative refinement routines FO4AFF and FO4AHF have so far been fruitless, bearing out the fact that double precision is costly on the CRAY.

In general, removing double precision inner products gives a speed up of 6 to 9 on a matrix of order about 100, and using SAXPY type code a further factor of the same order, while other changes give perhaps 10% further improvement in the speed up factor. Though only a score or so of routines have been fully worked, it is worth pointing out that they include the most used routines, and that they are called by other routines as auxiliaries so that their improvement has some effect on other routines' performances.

To give a measure of the absolute performance of the routines we convert the times to rates in megaflops. The original codes gave speeds often less than one megaflop. The modified codes give rates of roughly 20 megaflops (higher in the case of routines which perform complex arithmetic), which is fairly typical for FORTRAN. These figures are of course well below the 100-plus megaflops of the best CAL-coded routines.

REFERENCES

CRAY FORTRAN (CFT) Manual (1981) Cray Research, Inc. (Ref. SR-0009).

Higbie, L. (1979) Vectorization and Conversion of FORTRAN Programs for the CRAY-1 (CFT) Compiler, Cray Research, Inc. (Ref. 2240207).

Lawson, C.L., Hanson, R.J., Kincaid, D.R. and Krogh, F.T. (1979) Basic Linear Algebra Subprograms for Fortran Usage, *ACM Trans. Math. Software*, Vol. 5, no. 3, Sept. 1979, pp. 308-323.

Petersen, W.P. (1979) Basic Linear Algebra Subprograms for CFT Usage, Cray Research Inc.

Sydow, P.J. (1981) CRAY-1 Optimization Guide, Cray Research, Inc. (Ref. SN-0220).

APPENDIX

1. NAG Subroutine FO3AFF

```
      SUBROUTINE FO3AFF(N, EPS, A, IA, D1, ID, P, IFAIL)            FO3AFF.2
C     MARK 2 RELEASE. NAG COPYRIGHT 1982                            FO3AFF.3
C     MARK 3 REVISED.                                               FO3AFF.4
C     MARK 4.5 REVISED                                              FO3AFF.5
C                                                                   FO3AFF.6
C     UNSYMDET                                                      FO3AFF.7
C     THE UNSYMMETRIC MATRIX, A, IS STORED IN THE N*N ARRAY A(I,J), FO3AFF.8
C     I=1,N, J=1,N.  THE DECOMPOSITION A=LU, WHERE L IS A           FO3AFF.9
C     LOWER TRIANGULAR MATRIX AND U IS A UNIT UPPER TRIANGULAR      FO3AFF.10
C     MATRIX, IS PERFORMED AND OVERWRITTEN ON A, OMITTING THE UNIT  FO3AFF.11
C     DIAGONAL OF U.  A RECORD OF ANY INTERCHANGES MADE TO THE ROWS FO3AFF.12
C     OF A IS KEPT IN P(I), I=1,N, SUCH THAT THE I-TH ROW AND       FO3AFF.13
C     THE P(I)-TH ROW WERE INTERCHANGED AT THE I-TH STEP. THE       FO3AFF.14
C     DETERMINANT, D1 * 2.0**ID, OF A IS ALSO COMPUTED. THE         FO3AFF.15
C     SUBROUTINE                                                    FO3AFF.16
C     WILL FAIL IF A, MODIFIED BY THE ROUNDING ERRORS, IS SINGULAR  FO3AFF.17
C     OR ALMOST SINGULAR. SETS IFAIL = 0 IF SUCCESSFUL ELSE IFAIL = FO3AFF.18
C     1.                                                            FO3AFF.19
C     1ST DECEMBER 1971                                             FO3AFF.20
C                                                                   FO3AFF.21
      INTEGER ISAVE, IFAIL, IFAIL1, I, N, IA, ID, K, L, K1, K2,     FO3AFF.22
     * ISTART, J, P01AAF                                            FO3AFF.23
      DOUBLE PRECISION SRNAME                                       FO3AFF.24
      REAL Y, D2, D1, X, EPS, A(IA,N), P(N)                         FO3AFF.25
      DATA SRNAME /8H FO3AFF /                                      FO3AFF.26
      ISAVE = IFAIL                                                 FO3AFF.27
      IFAIL1 = 0                                                    FO3AFF.28
      DO 20 I=1,N                                                   FO3AFF.29
         CALL X03AAF(A(I,1), IA*N+1-I, A(I,1), IA*N+1-I, N, IA, IA, FO3AFF.30
     *    0.0, 0.0, Y, D2, .TRUE., IFAIL1)                          FO3AFF.31
```

```
      IF (Y.LE.O.O) GO TO 200                                   FO3AFF.32
      P(I) = 1.O/SQRT(Y)                                        FO3AFF.33
 20 CONTINUE                                                    FO3AFF.34
    Dl = 1.O                                                    FO3AFF.35
    ID = O                                                      FO3AFF.36
    DO 180 K=1,N                                                FO3AFF.37
      L = K                                                     FO3AFF.38
      X = O.O                                                   FO3AFF.39
      Kl = K - 1                                                FO3AFF.40
      K2 = K + 1                                                FO3AFF.41
      ISTART = K                                                FO3AFF.42
      DO 40 I=ISTART,N                                          FO3AFF.43
         CALL XO3AAF(A(I,1), N*IA+1-1, A(1,K), (N-K+1)+IA, Kl,  FO3AFF.44
*         IA, 1, -A(I,K), O.O, Y, D2, .TRUE., IFAIL1)           FO3AFF.45
         A(I,K) = -Y                                            FO3AFF.46
         Y = ABS(Y*P(I))                                        FO3AFF.47
         IF (Y.LE.X) GO TO 40                                   FO3AFF.48
         X = Y                                                  FO3AFF.49
         L = I                                                  FO3AFF.50
 40   CONTINUE                                                  FO3AFF.51
      IF (L.EQ.K) GO TO 80                                      FO3AFF.52
      Dl = -Dl                                                  FO3AFF.53
      DO 60 J=1,N                                               FO3AFF.54
         Y = A(K,J)                                             FO3AFF.55
         A(K,J) = A(L,J)                                        FO3AFF.56
         A(L,J) = Y                                             FO3AFF.57
 60   CONTINUE                                                  FO3AFF.58
      P(L) = P(K)                                               FO3AFF.59
 80   P(K) = L                                                  FO3AFF.60
      Dl = Dl+A(K,K)                                            FO3AFF.61
      IF (X.LT.8.O*EPS) GO TO 200                               FO3AFF.62
100   IF (ABS(Dl).LT.1.O) GO TO 120                             FO3AFF.63
      Dl = Dl*O.0625                                            FO3AFF.64
      ID = ID + 4                                               FO3AFF.65
      GO TO 100                                                 FO3AFF.66
```

```
120    IF (ABS(D1).GE.0.0625) GO TO 140                          F03AFF67
       D1 = D1*16.0                                               F03AFF68
       ID = ID - 4                                                F03AFF69
       GO TO 120                                                  F03AFF70
140    X = -1.0/A(K,K)                                            F03AFF71
       IF (K2.GT.N) GO TO 180                                     F03AFF72
       DO 160 J=K2,N                                              F03AFF73
          CALL X03AAF(A(K,1), N*IA+1-K, A(1,J), (N-J+1)*IA, K1,   F03AFF74
   *        IA, 1, -A(K,J), 0.0, Y, D2, .TRUE., IFAIL1)           F03AFF75
          A(K,J) = X*Y                                            F03AFF76
160    CONTINUE                                                   F03AFF77
180 CONTINUE                                                      F03AFF78
    IFAIL = 0                                                     F03AFF79
    RETURN                                                        F03AFF80
200 IFAIL = P01AAF(ISAVE,1,SRNAME)                                F03AFF81
    RETURN                                                        F03AFF82
    END                                                           F03AFF83
```

2. CRAY-1 Subroutine FO3AFF

```
      SUBROUTINE FO3AFF(N, EPS, A, IA, D1, ID, P, IFAIL)              FO3AFF.2
C     MARK 2 RELEASE. NAG COPYRIGHT 1972                              FO3AFF.3
C     MARK 3 REVISED.                                                 FO3AFF.4
C     MARK 4.5 REVISED                                                FO3AFF.5
C                                                                   F3AMOD48.1
C     CRAY-1 OPTIMISED IMPLEMENTATION.  AUGUST, 1982.                    ID1.2
C                                                                   F3AMOD48.3
C     CALLS THE FOLLOWING BASIC LINEAR ALGEBRA SUBROUTINES FROM SCILIB.  LIBID.2
C     (DETAILS ARE IN CRI PUBLICATION SR 0014 LIBRARY REFERENCE MANUAL)  LIBID.3
C     SNRM2  (EUCLIDEAN NORM OF REAL VECTOR)                        SNRM2ID.2
C                                                                     FO3AFF.6
C     UNSYMDET                                                        FO3AFF.7
C     THE UNSYMMETRIC MATRIX, A, IS STORED IN THE N*N ARRAY A(I,J),   FO3AFF.8
C     I=1,N, J=1,N. THE DECOMPOSITION A=LU, WHERE L IS A              FO3AFF.9
C     LOWER TRIANGULAR MATRIX AND U IS A UNIT UPPER TRIANGULAR        FO3AFF.10
C     MATRIX, IS PERFORMED AND OVERWRITTEN ON A, OMITTING THE UNIT    FO3AFF.11
C     DIAGONAL OF U. A RECORD OF ANY INTERCHANGES MADE TO THE ROWS    FO3AFF.12
C     OF A IS KEPT IN P(I), I=1,N, SUCH THAT THE I-TH ROW AND         FO3AFF.13
C     THE P(I)-TH ROW WERE INTERCHANGED AT THE I-TH STEP. THE         FO3AFF.14
C     DETERMINANT, D1 * 2.0**ID, OF A IS ALSO COMPUTED. THE           FO3AFF.15
C     SUBROUTINE WILL FAIL IF A, MODIFIED BY THE ROUNDING ERRORS, IS F3AMOD48.6
C     SINGULAR OR ALMOST SO. SETS IFAIL=0 IF SUCCESSFUL ELSE IFAIL=1. F3AMOD48.7
C     1ST DECEMBER 1971                                               FO3AFF.20
C                                                                     FO3AFF.21
      INTEGER ISAVE, IFAIL, I, N, IA, ID, J, K, K2, L, P01AAF       F3AMOD48.8
      DOUBLE PRECISION SRNAME                                         FO3AFF.24
      REAL D1, EPS, X, Y, A(IA,N), P(N)                             F3AMOD48.9
      DATA SRNAME /8H FO3AFF /                                        FO3AFF.26
C                                                                   F3AMOD48.10
      ISAVE = IFAIL                                                   FO3AFF.27
C     CALCULATE 2-NORMS                                             F3AMOD48.11
      DO 20 I=1,N                                                     FO3AFF.29
```

```
         P(I) = SNRM2(N, A(I,1), IA)                                    F3AMOD48.12
         IF (P(I).LE.O.O) GO TO 200                                     F3AMOD48.13
         P(I) = 1.O/P(I)                                                F3AMOD48.14
   20 CONTINUE                                                             FO3AFF.34
      D1 = 1.O                                                             FO3AFF.35
      ID = O                                                               FO3AFF.36
      DO 180 K=1,N                                                         FO3AFF.37
C     SEARCH FOR PIVOT ROW                                              F3AMOD48.15
         L = K                                                             FO3AFF.38
         X = ABS(A(K,K)*P(K))                                           F3AMOD48.16
         K2 = K + 1                                                        FO3AFF.41
         DO 40 1=K2,N                                                   F3AMOD48.17
            Y = ABS(A(1,K)*P(I))                                        F3AMOD48.18
            IF (Y.LE.X) GO TO 40                                           FO3AFF.48
            X = Y                                                          FO3AFF.49
            L = I                                                          FO3AFF.50
   40    CONTINUE                                                          FO3AFF.51
         IF (L.EQ.K) GO TO 80                                              FO3AFF.52
         D1 = -D1                                                          FO3AFF.53
         DO 60 J=1,N                                                       FO3AFF.54
            Y = A(K,J)                                                     FO3AFF.55
            A(K,J) = A(L,J)                                                FO3AFF.56
            A(L,J) = Y                                                     FO3AFF.57
   60    CONTINUE                                                          FO3AFF.58
         P(L) = P(K)                                                       FO3AFF.59
   80    P(K) = L                                                          FO3AFF.60
         D1 = D1*A(K,K)                                                    FO3AFF.61
         X = 1.O/A(K,K)                                                 F3AMOD48.19
         DO 160 J=K2,N                                                     FO3AFF.73
            A(K,J) = A(K,J)*X                                           F3AMOD48.20
CDIR*    IVDEP                                                          F3AMOD48.21
            DO 150 I=K2,N                                               F3AMOD48.22
               A(I,J) = A(I,J)  - A(I,K)*A(K,J)                         F3AMOD48.23
```

```
  150         CONTINUE                                          F3AMOD48.24
  160      CONTINUE                                               FO3AFF.77
  180 CONTINUE                                                    FO3AFF.78
      CALL FO3AEZ (Dl, ID)                                      F3AMOD48.25
      IFAIL = O                                                   FO3AFF.79
      RETURN                                                      FO3AFF.80
C                                                               F3AMOD48.26
  200 IFAIL = PO1AAF(ISAVE,1,SRNAME)                              FO3AFF.81
      RETURN                                                      FO3AFF.82
      END                                                         FO3AFF.83
```

3. Auxiliary Subroutine for the CRAY-1 Subroutines FO3AEF, FO3AFF and FO3AGF

```
      SUBROUTINE FO3AEZ (Dl, ID)                                   FO3AEZ.1
C                                                                  FO3AEZ.2
C     GIVEN A NORMALISED REAL NUMBER Dl, THIS ROUTINE              FO3AEZ.3
C     RETURNS Dl and ID SUCH THAT GIVEN Dl = Dl*(2**ID) WITH       FO3AEZ.4
C     1.O>Dl>=O.O625.                                              FO3AEZ.5
C     ASSUMES CRAY FLOATING POINT FORMAT OF SINGLE SIGN BIT,       FO3AEZ.6
C     15 BIT EXPONENT WITH BIAS OF 4OOOO OCTAL, AND 48 BIT MANTISSA.  FO3AEZ.7
C                                                                  FO3AEZ.8
C     WRITTEN FOR NAG LIBRARY AS AUXILIARY TO FO3AEF, FO3AFF AND FO3AGF.  FO3AEZ.9
C     SEPTEMBER, 1981                                              FO3AEZ.1O
C                                                                  FO3AEZ.11
      REAL Dl                                                      FO3AEZ.12
      INTEGER ID, MASKl                                            FO3AEZ.13
      DATA MASKl /O77777OOOOOOOOOOOOOOOOB/                         FO3AEZ.14
C                                                                  FO3AEZ.15
      IF (Dl.NE.O.O) GO TO 2O                                      FO3AEZ.16
                                                                   FO3AEZ.17
      ID =O                                                        FO3AEZ.18
      RETURN                                                       FO3AEZ.19
C                                                                  FO3AEZ.2O
C     EXTRACT EXPONENT AND SHIFT TO GIVE INTEGER VALUE             FO3AEZ.21
   2O ID = SHIFTR(Dl.AND.MASKl,48)                                 FO3AEZ.22
C     SUBTRACT WEIGHTING AND DISCARD LEAST SIGNIFICANT BITS        FO3AEZ.23
      ID = (ID - 37775B). AND..NOT.3B                              FO3AEZ.24
C     SCALE ORIGINAL VALUE FOR NEW EXPONENT                        FO3AEZ.25
      Dl = Dl*(2.O**(-ID))                                         FO3AEZ.26
      RETURN                                                       FO3AEZ.27
      END                                                          FO3AEZ.28
```

TOUGH PROBLEMS IN REACTOR DESIGN

W.D. Collier, C.W.J. McCallien and J.A. Enderby

(Central Computer Services, UKAEA, Risley)

1. INTRODUCTION

The UKAEA Northern Division deals with physics and reactor design problems in collaboration with other UKAEA establishments at Harwell, Winfrith and Culham, with CEGB who are responsible for nuclear power stations already running, NNC who are responsible for particular designs of reactors being built and with BNFL who deal with fuel fabrication, transport and reprocessing.

This means that a very wide range of problems is tackled, questions of what goes on deep inside the reactor core, of the integrity of fuel pins and core components, study of stresses in structural components, heat transfer and fluid flow both in the reactor and in the non-nuclear parts of the generating plant, design of transport containers for new and used fuel and so on. Similar calculations are performed for a series of reactor types, Magnox, AGR, SGHWR, Fast Reactor and now PWR, as well as for fusion research.

This implies an immense need for computational facilities. This computing activity is generated from all UKAEA and BNFL sites from Dounreay on the north coast of Scotland where the main interest is the fast reactor down to Winfrith on the Dorset coast where the SGHWR is situated.

Users at the northern sites access the dual ICL 2982 at Risley directly and the IBM 3033 at Harwell via the Risley communications centre. Although the CRAY is attached to the 3033, software has been developed to allow northern division users to access it as if it were a back end processor to the 2982.

From the Authority's very wide range of computing interests three topics have been selected:

(i) solution of neutronics problems by diffusion theory using finite differences

(ii) solution of neutronics problems by transport theory using Monte Carlo methods

(iii) structural problems using finite elements.

2. FINITE DIFFERENCES

During the design and operation of a nuclear reactor, the neutron flux must be determined as a prelude to calculating the power output and temperatures in individual components such as fuel pins. An adequate approximation to the flux can often be obtained by solving the multi-group diffusion equations.

$$-D_g \nabla^2 \phi_g + \Sigma_g^{rem} \phi_g = \sum_h \sigma_{h \to g} \phi_h + \frac{1}{k} \sum_h \chi_g \nu \Sigma_h^f \phi_h$$

where there is one equation for each energy group g $(1 \le g \le G)$, and where

ϕ_g is the unknown flux or eigen-vector

k is the unknown eigen-value (the only scalar in the equation)

D_g is the diffusion coefficient

Σ_g^{rem} is the total removal cross-section with contributions from capture and out-scatter

$\sigma_{h \to g}$ is the scatter cross-section out of group h into group g

χ_g is the emergence spectrum

$\nu\Sigma_h^f$ is the average number of neutrons born in one fission event times the fission cross-section.

It is seldom possible to assume the last five items are constants because in a reactor, the isotopic concentrations can change markedly from place to place. They are usually assumed to be functions of the regions and a region can be as small as a single mesh box.

The finite difference approximation to these equations can be written in matrix notation

$$M\phi = \frac{1}{k} F\phi$$

This is an eigen-problem usually solved by a set of "outer" iterations in which the well known matrix powering algorithm is applied to determine the largest value of k and the corresponding flux vector ϕ (Wachpress (1966)). The outer iterations can be accelerated by Chebyshev extrapolation, Aitken's δ^2 method, or coarse mesh rebalancing (Wachpress (1966), Hageman (1963) and McCallien (1976)). Each outer iteration requires the solution of

$$M\phi^{(n+1)} = F\phi^{(n)}$$

to obtain $\phi^{(n+1)}$. This is usually achieved by point or block successive over-relaxation (SOR) of a further "inner" set of iterations (Wachpress (1966), Hageman (1963) and McCallien (1970)).

A typical calculation for the prototype fast reactor (PFR) has 1450 mesh points on a 2-D triangular mesh which, with six energy groups, gives 8700 unknowns. This is well within the capacity of the ICL 2982 taking about a minute. To determine the variation in the Z direction requires the introduction of some 30 mesh intervals vertically making the problem much larger. A detailed knowledge of the spectrum might require the use of 37 energy groups rather than six, and the same calculation for a commercial reactor would have twice as many (x,y) points. Consequently such a problem would have over 3,000,000 unknowns requiring more than ten hours of processor time, and perhaps 40 elapsed hours allowing for disc transfers.

At first sight, the main difficulties in solving the equations appear to be numerical ones associated with Chebyshev extrapolation and SOR. However there is also a very real data handling problem. Scatter is represented by a matrix $\sigma_{h\to g}$ which can have order (10x37). Thus there can be as many as twenty coefficients to be stored for every unknown. If these make the problem too large for the main memory, they have to be held in secondary storage and swapped in and out adding to the cost and increasing the elapsed time. There are several ways in which such data transfers can be reduced in number. During SOR of the inner iterations disc transfers can be reduced by the use of concurrent iterations, (Wachpress (1966) and Curtis (1962)) and during both the outer and inner iterations transfers can be reduced by periodic coarse mesh rebalancing in which the solution of an ancillary coarse mesh problem is periodically used to reshape the latest estimate of ϕ (Wachpress (1966), McCallien (1976) and Froelich (1969)). The coarse mesh should be carefully chosen to be sufficiently coarse for all its coefficients to fit into main memory thereby avoiding repeated disc transfers.

The use of a super-computer such as a CRAY-I with a processor five or six times faster than the ICL 2982 is an attractive possibility for large reactor calculations. The calculations described here, if based on a line variation of SOR, would lend themselves to optimisation to take advantage of vector arithmetic in the CRAY giving a further factor of perhaps fifteen in speed. Thus the processor time for solving the problems with 3,000,000 unknowns might be reduced from ten hours to seven minutes! Unfortunately, speeding up the central processor does not tackle the enormous data handling problem, and at this stage, without practical experience on the CRAY, it seems unlikely that it will be possible to feed coefficients from secondary storage into main store fast enough to keep the vector registers in continuous operation. Thus the elapsed time is likely to remain considerably in excess of the computational time.

3. MONTE-CARLO

The interior of a nuclear reactor is geometrically very complex and when 3-D calculations need to be performed, considerable simplifications must be made to produce finite difference or finite element computer models which are feasible to run. I have already mentioned the running times required for diffusion calculations. Transport theory can sometimes be totally out of the question.

Monte-Carlo methods provide a 3-D description of the item under consideration, generally by building it up from elementary shapes, spheres, cylinders and cuboids, to as great a degree of complexity as desired. UKAEA's MONK (Sherriffs (1978)) aids the user by giving cross-section pictures of the model on a colour graphics screen. The model is not provided with a mesh. Instead, individual neutrons are followed through the model from birth or entry into the system until they are absorbed or escape from the system. During its lifetime each neutron undergoes a number of interactions with nuclei, namely elastic scatter, inelastic scatter, absorption, and absorption prior to fission of the nucleus. The length of path before the interaction, which interaction occurs, and the energy and directions of any emergent neutrons are determined by a random number generator which of course gives the method its name.

Many orders of magnitude fewer neutrons can of course be followed than actually exist in the real system, nevertheless very large numbers of neutrons must be followed if the variance of the results is to be low. In a simple system, around 10,000 neutrons may be expected to give a standard error of 1% on reactivity (k of the previous section).

An initial population of neutrons is generated (in practice in batches of around 100 at a time rather than 10,000 at once) whose distribution with regard to position energy and direction is, for a reactivity calculation, an approximation to the solution. Individual neutrons from the population are tracked through the model until they are absorbed, lost from the system, or cause a fission. Daughter neutrons from the fission become the population for the next stage. At each stage, the normalised initial distribution more closely resembles the eigen-solution sought. Criticality (k) is determined by comparison of the numbers of neutrons in successive stages.

For an imposed source (e.g. shielding) problem the initial population corresponds to the imposed source. Any daughter neutrons are added to the current population and the calculation requires only a single stage.

Neutron currents may be obtained by counting neutrons which pass through a specified boundary.

The calculations required to track a neutron are in some cases quite complex. To check in 3-D where a path intersects the boundary of a region of complex shape is itself a difficult exercise in 3-D coordinate geometry. The total number of neutrons followed can be very large indeed and immense amounts of computer time can be consumed by Monte-Carlo calculations. The speedup factors available even with only a small amount of optimisation from certain super-computers will considerably increase the range of problems which can be tackled.

Parallel processors suggest the possibility of following many neutrons in parallel. Unfortunately their histories do not remain parallel. Once an interaction has occurred they follow increasingly divergent paths. It is not clear that the present algorithms can be altered to find large scale parallelism. This may be possible with completely restructured methods and all suggestions are welcome.

4. FINITE ELEMENTS

4.1 The Risley system UNCLE

Experience with finite elements at Risley goes back about fifteen years. For about the last six, we have been developing a very general finite element scheme called UNCLE (Enderby (1979)) and its associated modules for heat transfer, stress analysis in continua and stress analysis in shells, frameworks and systems of pipes (Johnson (1981), Richardson (1982) and Knowles (1981)).

A particular feature of UNCLE is the way it provides the user with a powerful and concise yet natural methodology for describing structures. Those who have tried finite element modelling in 3-D will appreciate the difficulties involved and the need for suitable tools.

A particular feature of the heat transfer module is its method of dealing with radiation between different surfaces of a model whatever their relative positions, automatically calculating view-factors and making allowance for obstructions.

Further discussion of UNCLE itself is inappropriate at this point except to note that the provision of these powerful facilities has allowed users to produce with relative ease models which they would not have considered feasible a short while ago and which in some cases stretch our present computing capabilities to the limit.

4.2 A stress analysis investigation

Consideration of a specific case will show why finite element analysis also provides "tough" problems. Consider a drum with many nozzles subject to internal pressure and also to a severe thermal transient causing substantial stress in the weld region of one of the nozzles. Now postulate a flaw in the weld or in the nozzle casting. What effect does this have on the performance? Is the stress intensity such that the flaws will grow or not? Is growth stable or unstable? If stable growth is expected, is an ultra-sonic inspection every year sufficient to avoid failure?

A half section of the drum was taken with an end cap and a single nozzle (Figs. 1, 2). The number of nodes generated was 2941 giving 8823 unknowns in the stress analysis calculation. In some contexts this would not be regarded as a large number of equations but as is usual in finite element analysis the matrix of coefficients is not well structured and even with optimal ordering the bandwidth is quite large.

The number of floating point operations required approximates to

$$2nw^2 - 4w^3/3$$

where n is the number of equations and w is the bandwidth. The bandwidth for this problem is about 300 for the first 3000 equations then it increases considerably. The whole case including post-processing to obtain stresses took 13000 work units, that is around 10,000 seconds or three hours of ocp time over an elapsed time of eight hours. The amount of peripheral storage used is not known precisely but is in the region of 50-60 Mb. Relative to the available resources this is regarded as a fairly large job.

It was clearly impractical to introduce detailed representation of the postulated flaws into this model so a simplified model of the flaw region was defined, and stresses obtained from the full model were imposed on the flaws. The buried flaws are radial regions inserted into the essentially rectangular block. Crack-tip elements are then inserted all around the edge of each flaw (Fig. 3). This model very clearly illustrates the first area in which these problems

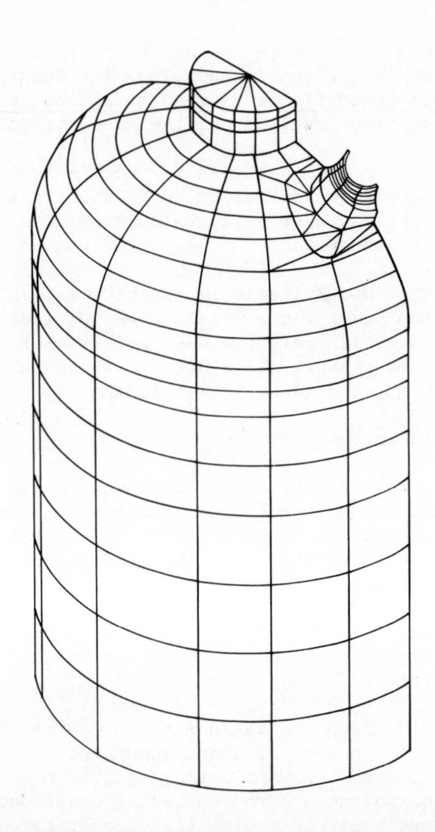

Fig. 1 Hidden Line Plot of Drum and Nozzle

are regarded as tough. The mental effort involved in developing the complex mesh inside a region into which you cannot see is immense (Figs. 4,5).

The total number of nodes generated was 6288 giving 18864 unknowns. The band-width varies from 700 to 4500 but is around 1500 over large regions. Assuming this value everywhere we find that the solution of the equations requires around 8×10^{10} floating point operations or about 22 hours ocp time on the 2982. This would reduce to 13 minutes on a 100 Mflops machine.

This underlines the second reason for regarding these problems as tough. The calculations facing us demand, and the systems we have already designed allow us to set up, problems so large that we do not have the computer power to solve them.

4.3 What can we speed up?

4.3.1 Data input is essentially a serial process. There will be scope for local vectorisation in do loops, but no obvious large scale

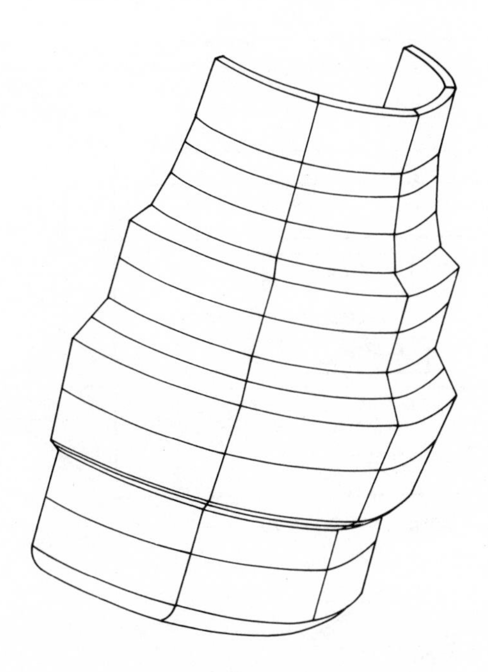

Fig. 2 Hidden Line Plot of Nozzle

parallelism. In any case the length of time spent in this part of the program is not great.

4.3.2 Setting up the element equations and assembling the global matrix can be time consuming though in large cases it is almost always dominated by the solution time. However we might look in five areas for parallelism.

(i) small scale parallelism in do loops

(ii) rewriting sorts. These are obvious candidates for parallel processing by vector and array processors, but the serial sort in UNCLE is already very fast.

(iii) calculating equations for many elements together. At present we do not see this as a real possibility. A model may consist of a wide range of very different elements and particularly in non-linear analysis apparently similar elements may require different treatments in different circumstances.

(iv) Element equations. There may be considerable scope for optimising coordinate transformations and other matrix operations for a given high speed processor.

(v) Solving the global equations. This is the most time consuming part of the program and the one which shows most promise for optimisation.

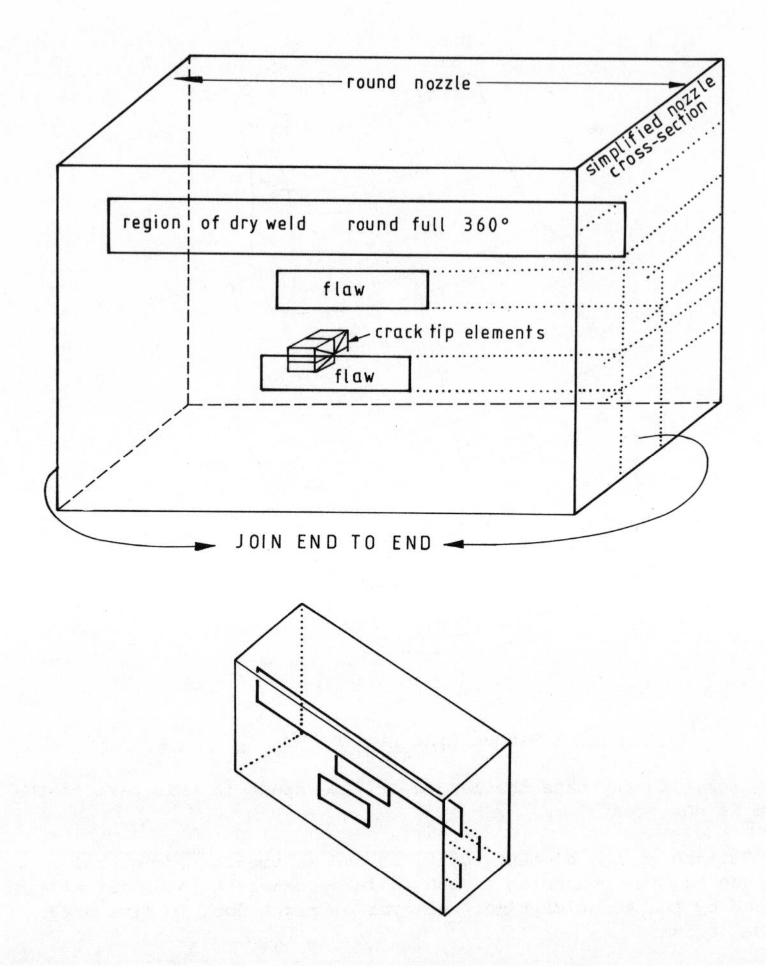

Fig. 3 Model Using Crack Tip Elements

4.3.3 The classical direct methods remain the only viable means of solving the sort of equations produced by finite element methods. The choice between them is largely a matter of taste and convenience since they are algebraically equivalent.

The frontal method is recommended if it is not convenient to assemble the full global matrix but it offers nothing extra in the way of computational efficiency over straightforward Gauss elimination.

UNCLE assembles the global matrix efficiently and so the speed of solution depends on selecting an algorithm which deals suitably with the sparseness of the equations and which manages the large volume of data involved with the least possible number of disc transfers. Many methods have been suggested. The subroutine SPARSE at Risley has the particular advantage that it is not restricted by band-width, nor do

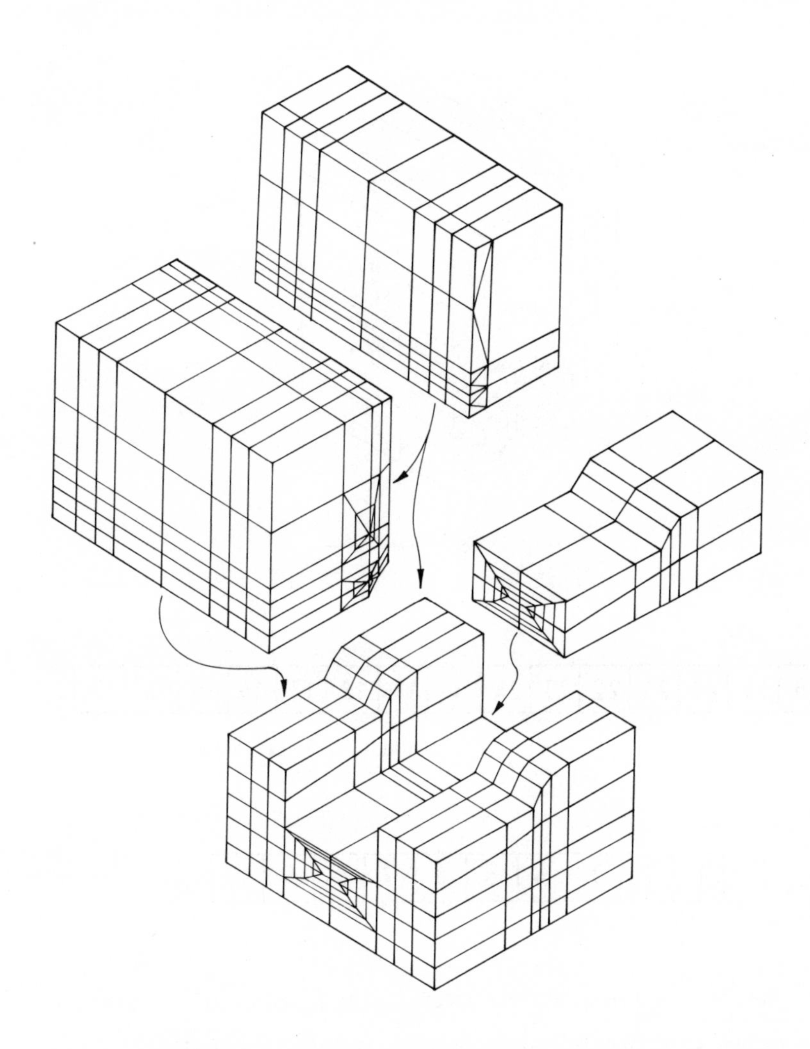

Fig. 4 Stages in Construction of Model

users need to select "in-core" or "out-of-core" versions. A version of this algorithm is being developed on the DAP.

5. CONVERSION OF THE SPARSE EQUATION SOLVER TO THE DAP

The method chosen to solve the global equations was constrained by certain conditions. Only for a range of small cases will sufficient fast store be available for the matrix and all intermediate items to coexist. Efficient queueing and backing store systems are required. It should be possible to solve for a number of right hand sides (load vectors) simultaneously. The processed matrix should be able to be saved for later application to other right hand sides.

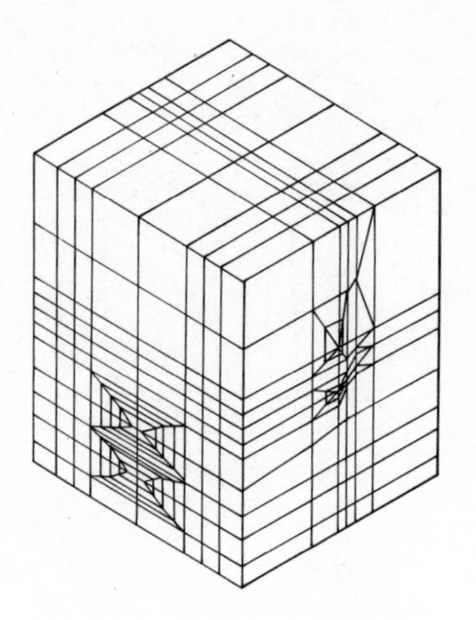

Fig. 5 Completed Model

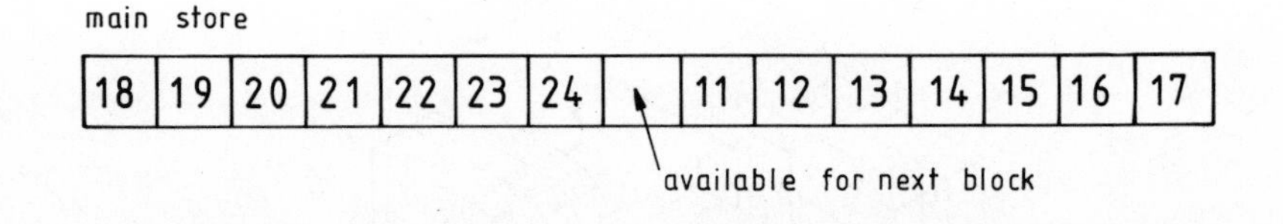

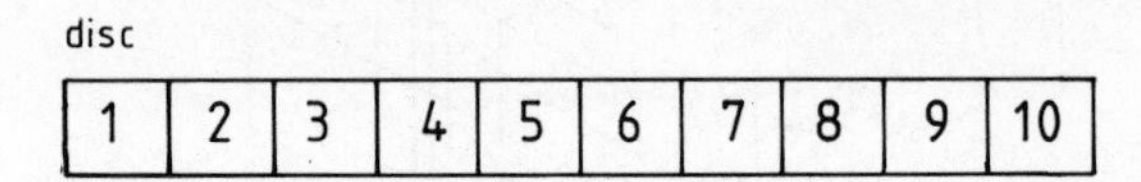

Fig. 6 Positions of Data Blocks in Main Store and on Disc

The method of column by column Gauss elimination used in the current version of SPARSE seems to carry over into submatrix form without problems. The matrix is broken into equal blocks (of size 64x64 for the present DAP) then in the algebra all scalar matrix elements are replaced by submatrices.

If fewer than 64 equations were to be solved consideration would be given instead to Gauss-Jordan since the eliminations above the diagonal are done in parallel with those below and so, on the DAP, are effectively done for nothing. However a block form of Gauss-Jordan is more expensive than a block form of normal Gaussian elimination since the elimination of blocks above the diagonal is as costly as the elimination of those below whereas the back substitution process is an order of magnitude cheaper.

The first important characteristic of the SPARSE method is that the matrix is stored column by column and no zero item is ever stored. Gaussian elimination is regarded as a row procedure since at stage k in order to eliminate all the items in column k with $i > k$ suitable multiples of row k are subtracted from all rows below row k. If however we only consider column k, the appropriate multipliers may be obtained and stored and column k discarded. If column $k + 1$ is then brought into main store, all the previously stored multipliers may be used to subtract appropriate multiples of the above-diagonal items in column $k + 1$ from lower rows. The multipliers needed to eliminate the below diagonal items of column $k + 1$ may then be calculated and stored, and the process repeated for column $k + 2$. Thus we never need to have more than a single column of the matrix in store at once. At each stage, the non-zero items remaining above the diagonal after the operations have been performed on the column, are stored in a second queue for use in the back substitution.

The second important characteristic of the method is that only the necessary operations are performed on each column. If the first few items in the column are zero then all operations which would subtract them from lower rows will have no effect. If the first non-zero item in the current column is in row i then the first operations which need to be considered are those produced to eliminate the items below the diagonal on column i. This of course saves unnecessary work (as is essential with any sparse matrix solver) but it has a much more fundamental effect on the strategy for storing the operations.

A region of main memory as large as possible is set aside for the storage of the operations. If the number of equations and band-width are modest, it may be that the whole calculation can progress without recourse to backing store. In general however, the main memory will be used cyclically, blocks being written to disc to make room for new ones in main store. At some stage, the blocks in main store and on disc will be as in the Fig. 6.

To do the elimination on the current column, we know to which column we must return to find the first necessary operation. It is a simple matter to keep a table to point to the appropriate block. Even with very large numbers of equations, if the band-width is reasonable, we will find the operations we require still in main store. Hence there is a wide range of cases in which the operations queue will be output to disc, but will not be read during the elimination phase.

If the band-width is locally large then certain blocks will need to be read back, but only those whose operations are required, and not the whole queue. If the band-width then reduces again, the program automatically reverts to writing but not needing to read blocks back.

As an example, consider a T joint (Fig. 7). The matrix band-width is six almost everywhere, but very locally it becomes 62 (nodes 55 and 117 are in the same element). If the numbering had been horizontal throughout and the numbering of the two pieces merged, the maximum band-width would only have been 21 but this would have been true for 100 nodes. A wavefront band-width optimiser would have produced a maximum band-width of 14 but the band-width would have been at least eleven for about 90 nodes in all. The original ordering is best in general and also illustrates the algorithm's action.

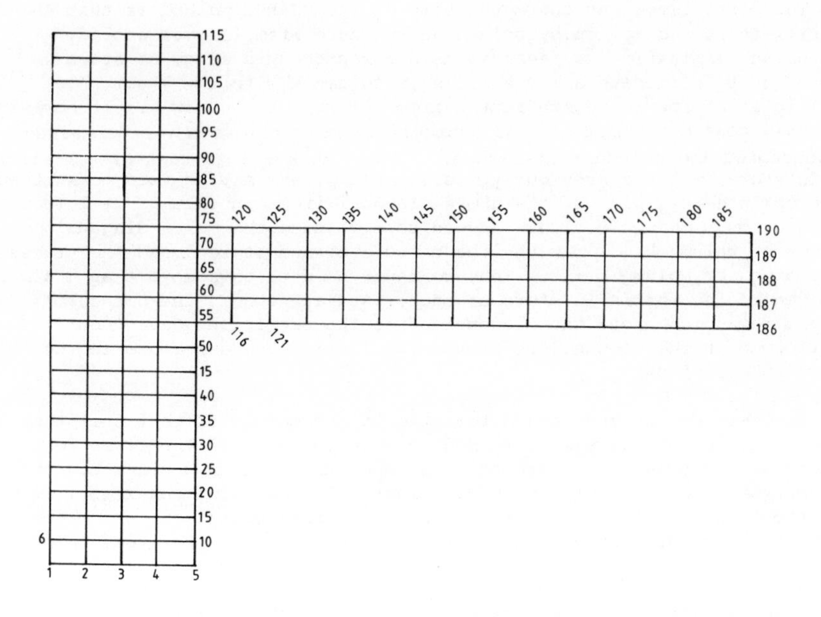

Fig. 7 The T Joint and Node Numbers

The matrix non-zero items are distributed as in the Fig. 8. For simplicity I will deal with individual matrix elements not with 64 x 64 blocks of elements. Right up to node 115 we need only consider elements generated up to six columns previously, these will be found in main store. At column 116 we find a non-zero item in row 55 therefore operations generated during the elimination of column 55 are relevant. These will need to be accessed from disc. Columns 117-120 are similar. From column 121 onwards we again see that only operations generated a maximum of six columns previously need to be considered and so no more reads from disc are required.

When the elimination is complete, the right-hand side vectors are likely to be mostly non-zero so the whole operations queue and back-substitution queue are read in and applied once. (If there are more than 64 right-hand sides then this phase will be completed a correspondingly greater number of times.

The usefulness of the algorithm depends on the amount of parallelism that can be utilised, and this clearly depends on the density of the non-zero blocks of the matrix. Since the number of adds and multiplies is about the same, and the times taken on the DAP are 180 μs and 285 μs respectively, the DAP should perform 4096/230 million, 18,000,000 instructions per second. This is a useful factor of twenty up on the 2982 but relies on using the DAP 100% efficiently. Timings of real 2982 runs using the scalar version of SPARSE against instruction counts in the DAP coding produced speedup factors in the range 2 - 8 with a few over 10 where it appeared that band-width considerations were working against the serial version without going outside the blocked band on the DAP version.

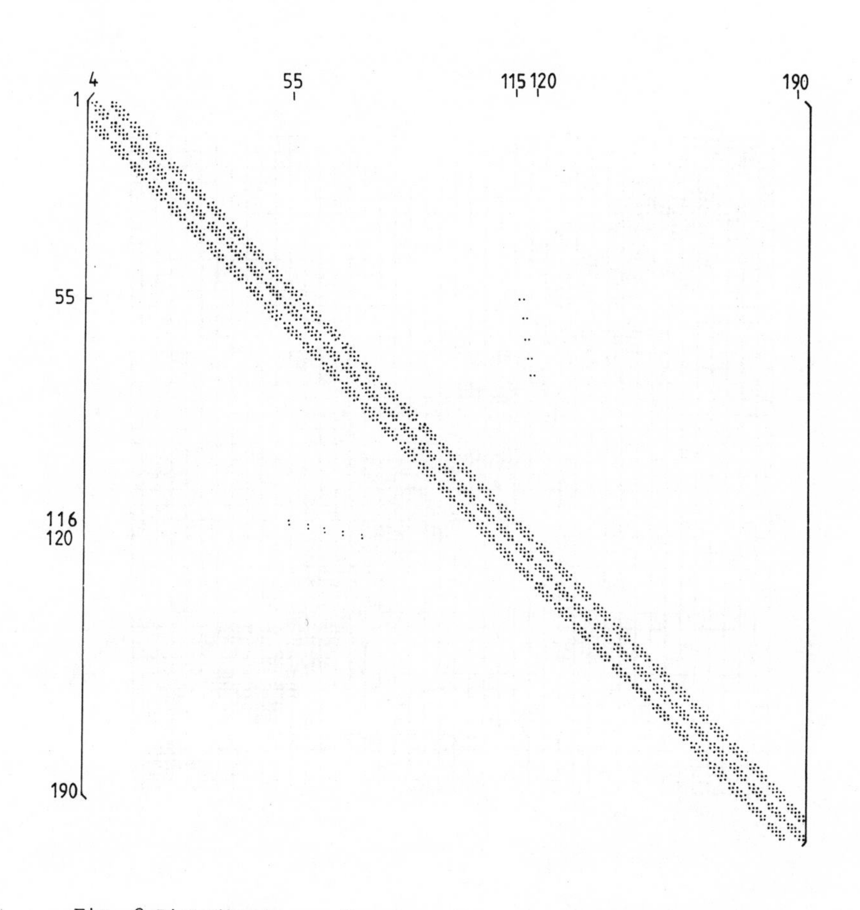

Fig. 8 Distribution on Non-Zero Items in the Global Matrix

Three dimensional stress problems give greater density than 2-D. With 20 node isoparametric bricks there may be as many as 240 non-zero items on a row. Radiation in heat transfer problems increases the connectivity of the problem considerably and hence also the matrix density.

The Figs. 9 - 11 show the distribution of non-zero items in the first 2733, 384, and 100 rows and columns of the drum and nozzle problem relative to 64 x 64 blocks. When the effects of "fill-in" are considered, it appears that the DAP will be used at around 50% efficiency, thus offering a factor of ten increase in speed over the 2982.

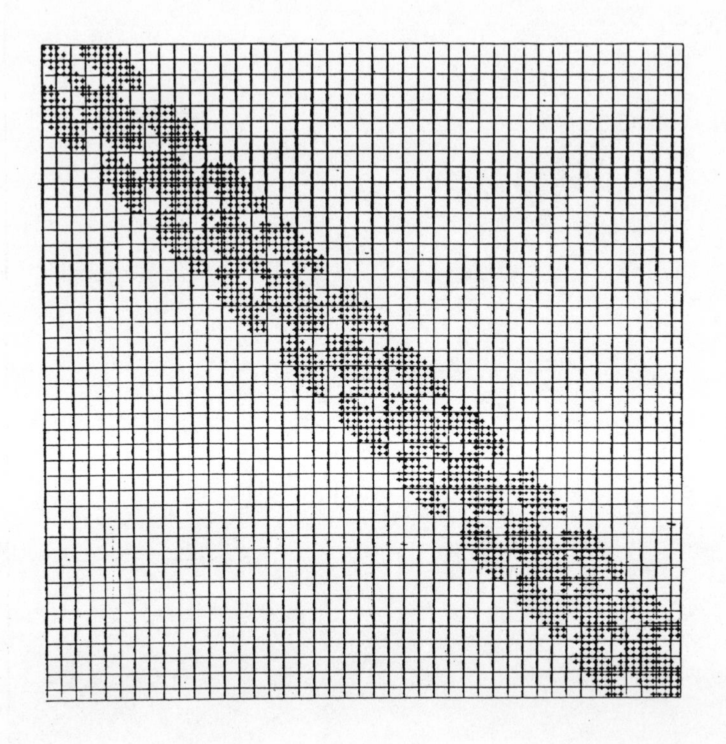

Fig. 9 Distribution of Non-Zero Items in Global Matrix (FIRST 2733 Columns)

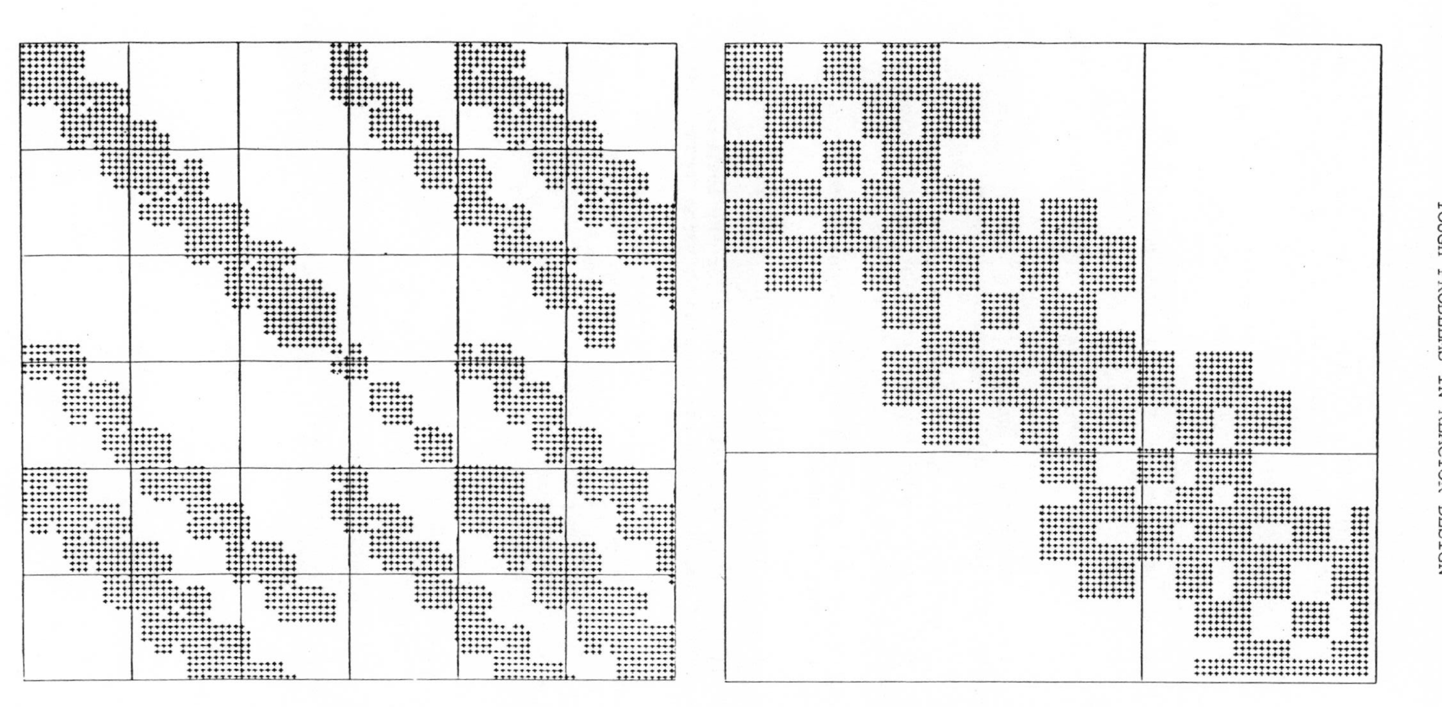

Fig. 10 Distribution of non-zero items in global matrix (first 384 columns)

Fig. 11 Distribution of non-zero items in global matrix (first 100 columns)

REFERENCES

Wachpress, E.L. (1966) Iterative solution of elliptic systems and applications to the neutron diffusion equations of reactor physics. Prentice Hall

Hageman, L.A. (1963) Numerical methods and techniques used in the 2D neutron diffusion program PDQ-5. WAPD-TM-364

McCallien, C.W.J. (1976) Computational Methods in Classical and Quantum Physics. (M.B. Hooper, Ed.) Advance Publications, pp. 231-260

McCallien, C.W.J. (1970) The solution of reactor diffusion problems. The Computer Journal, Vol 13, no 4, pp. 369-377

Curtis, A.R. (1962) Numerical solution of ordinary and partial differential equations. (L. Fox, Ed.) Pergamon Press, pp. 391-392

Froelich, R. (1969) Computer Independence of large reactor physics codes with reference to well balanced computer configurations in CONF 690401, pp. 451-470

Sherriffs, V.S.W. (1978) MONK - A general purpose Monte Carlo neutronics program. SRD-R-86

Enderby, J.A. (1979) An Introduction to UNCLE. ND-R-255 (R)

Johnson, D. (1981) TAU: A computer program for the analysis of temperature in 2 and 3 dimensional structures using UNCLE. ND-R-218(R)

Richardson, T. (1982) CAUSE: 2 and 3 dimensional stress analysis of continua using UNCLE. ND-R-503(R)

Knowles, J.A. (1981) FAUN: Analysis of frameworks pipework and shells using UNCLE. ND-R-569(R)

ON THE DESIGN AND IMPLEMENTATION OF A PACKAGE FOR SOLVING A CLASS OF PARTIAL DIFFERENTIAL EQUATIONS ON THE ICL DISTRIBUTED ARRAY PROCESSOR

S.L. Askew and F. Walkden

(Department of Mathematics, University of Salford)

SUMMARY

This paper begins with a description of problems in the field of Computational Fluid Dynamics which the package will solve. Suitable (explicit) methods of solution for use in conjunction with the package are summarised. General design features of the package are then described. Various difficulties encountered in the designing and development of the package are discussed, with the solution to these problems where they are known. Finally, a description of the current state of the implementation is given.

1. INTRODUCTION

The package described in this paper (PDEPACP) is a parallel version of an existing serial package (PDEPACA) written at Salford University. The latter was written as a research tool for, and by, the Computational Fluid Dynamics (CFD) research group, to reduce the burden of programming each research problem individually. However, some subsequent research, see Stansfield (1981), in which a number of computer programs designed to solve a selection of CFD problems were analysed and compared, yielded two interesting pieces of information. Firstly, the data structures used in each program were essentially very similar. Secondly, a large proportion of each program was 'housekeeping' code, again very similar, whilst that code which was particular to each program, i.e. specialised operations relevant only to the specific problem being solved, formed a relatively small part. The first of these results has led to a separate avenue of research to design suitable structures and handlers. The latter has confirmed the usefulness of a package like PDEPACA, to supply all the necessary 'housekeeping' code in a suitable generalised form to be used as a framework to which a user can add the specialised code relevant to his problem, thus greatly reducing the burden of programming effort.

The parallel version of the package, described here for the ICL Distributed Array Processor (DAP), has the additional aim of allowing the user, as far as possible, to translate his specialised serial code directly and simply into a parallel form, by virtue of creating data structures and routines which handle blocks of data, in parallel, in exactly the same manner as single points are handled in a serial program.

This paper is intended as a report of the ongoing work to develop PDEPACP.

2. DESCRIPTION OF SUITABLE PROBLEMS

The package is intended to solve real problems from the field of supersonic aerodynamics. We are concerned with solutions in both two and three dimensions: elliptic problems in two dimensions and hyperbolic (pure supersonic) in three dimensions. From the field of supersonic flow the simplest shapes considered as test problems would be flow past a wedge in 2D or a cone in 3D. At the other extreme, it is intended that real shapes such as that shown in Fig. 1 will be studied. In order to study such shapes, which are formed from discrete data defined by blue prints, two graphical packages (SURDATPAC and SOCSPAC) which produce the numerical representations of components and complete shapes and suitable solution grids, have been written by the CFD research group (Walkden, Law and Stansfield (1979)).

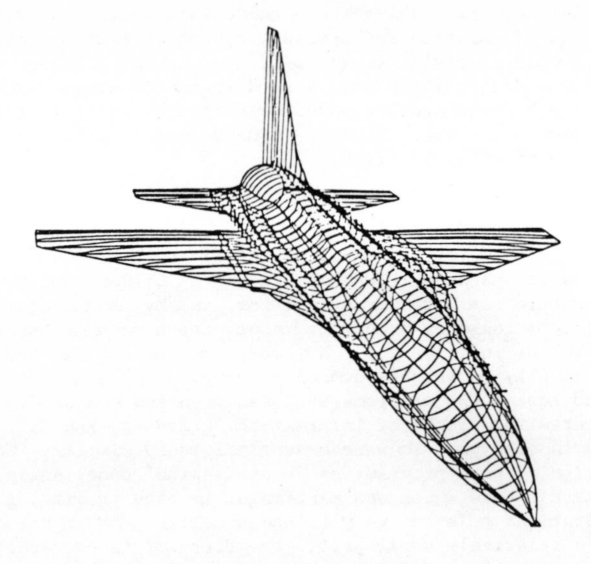

Fig. 1

3. METHODS OF SOLUTION

The package is intended for use with finite difference or finite volume algorithms in both two and three dimensions. The package assumes a rectangular coordinate system for its computational space. In practice the grid which the user chooses in cartesian space is unlikely to be rectangular, but rather to fit the physical boundaries of the problem; for example, the body shape below and shock position or free stream values above. In order to accommodate this, the package will accept user specified subroutines which define transformations between the two spaces.

PDEPACP is designed to use solution methods based on simple point relaxation schemes to solve the discretised problem, but, as explained in the introduction, considered not at individual points but operating on blocks of points, termed regions, in parallel. The efficient design and implementation of such methods for the DAP forms a further topic of research undertaken by the CFD research group.

4. DESCRIPTION OF PDEFACP

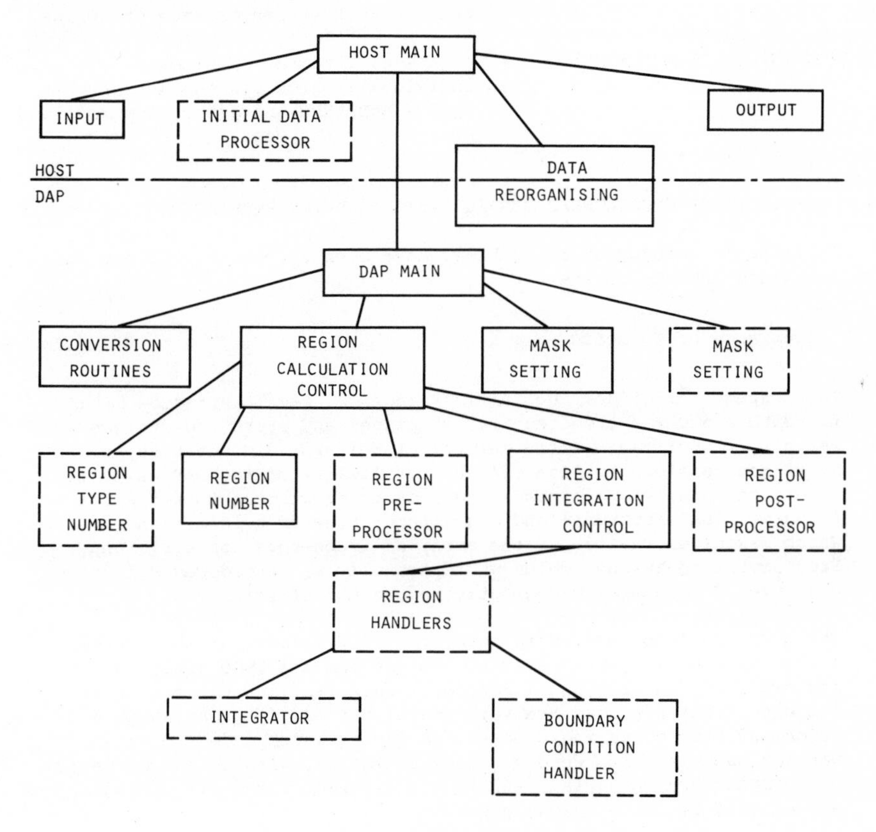

Fig. 2

The structure of the package is shown in Fig. 2. A solid outline depicts package routines, a broken outline user defined routines. A brief description of the function of major routines follows:-

HOST MAIN
- controls data input and output in a variety of forms.
- organises data storage.
- allows preprocessing of initial data by user.

DAP MAIN
- initiates standard data conversion. required for DAP-HOST interaction.
- controls set up of standard and user defined logical masks.
- controls the order in which regions are processed.

REGION CALCULATION CONTROL - controls calculations for each individual region.
- allows the user to pre- and post-process region data, e.g. post-process to test convergence criterion.

REGION INTEGRATION CONTROL - organises integration of each individual region via the appropriate REGION HANDLERS and associated routines provided by the user.

5. DIFFICULTIES ENCOUNTERED DURING DESIGN AND IMPLEMENTATION

Three major categories of problems have been encountered so far in implementing PDEPACP on the DAP.

5.1 Ignorance of Concepts

This category includes the initial stages of familiarisation with the 2980 and DAP - architecture, languages and operating systems - where the knowledge existed but we, at Salford, did not have it. It is not possible to 'ignore' the parallelism of the DAP as it is with some parallel machines. Code has to be written in a special language - DAP Fortran - and algorithms structured in a manner which maximises the use of the array of processors for efficiency. Programming techniques which would normally be considered effective, e.g. use of pointers, can greatly reduce DAP efficiency.

The 2980 operating system is completely different from the previous ICL (e.g. 1900 series) operating systems and the 2900 series has its own extended version of Fortran. Once familiar the facilities and extensions provided are very useful but familiarisation is a major and time consuming process. A factor which added to this was ICL manuals which were not self-contained, with a lack of basic, or introductory, material, and in our experience the effect is exacerbated by being remote users.

The answers to most problems in this section have come by trial and error, the additional documentation produced by QMCCC and frequent contact with the DAP Support Unit.

5.2 Design and Development

5.2.1 System Architecture/Nature of Problem - lack of perfect match

The solution methods which the package accommodates, i.e. finite difference and finite volume algorithms involve considerable calculations on, and manipulation of, matrix and vector data. The DAP with its associated version of Fortran is specifically designed to handle matrix and vector data. Since the problems we are concerned with are also large, the DAP should be a natural choice of machine. The match, however, is not perfect. Our problems are not necessarily large in the two dimensional sense (a $|64\times64|$ grid of processors) that the DAP is large. In practice $|30\times30|$ grids are quite adequate for our problems, which are 'big' in the sense that there are a large number of data items which are associated with each grid point in two or three dimensions and also large amounts of data specifying boundary shapes.

In Section 3 the package was described as operating on blocks of points, or regions, in parallel. These regions are not $|64\times 64|$, but are $|8\times 8|$, with the data stored in DAP Fortran vector mode. If we work in double precision on the DAP, vector operations utilise the full $|64\times 64|$ processor array. Fig. 3.1 shows the transformation of points in an $|8\times 8|$ array in computational space onto a DAP vector and Fig. 3.2 shows the storage of a set of such vectors, one vector for each dependent variable and a set of such vectors for each region, in a DAP vector array. Clearly, some points which are adjacent in the $|8\times 8|$ region of computational space are dispersed in vector store, and points adjacent in computational space but in separate regions are dispersed in the array.

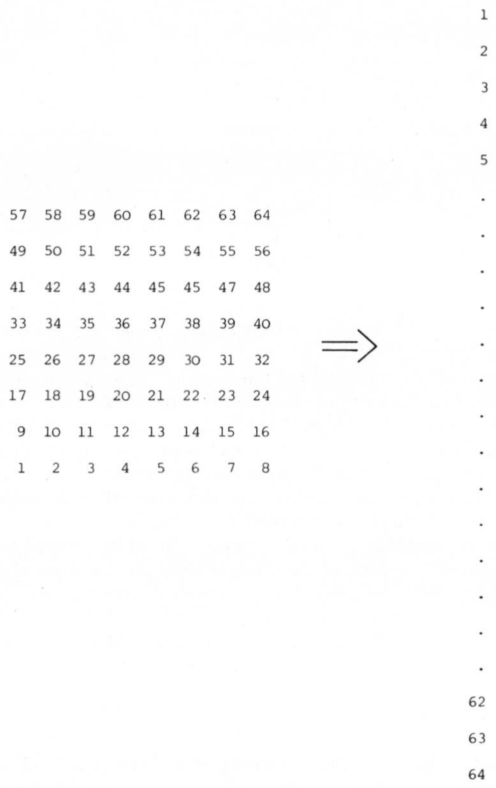

57	58	59	60	61	62	63	64
49	50	51	52	53	54	55	56
41	42	43	44	45	45	47	48
33	34	35	36	37	38	39	40
25	26	27	28	29	30	31	32
17	18	19	20	21	22	23	24
9	10	11	12	13	14	15	16
1	2	3	4	5	6	7	8

$\Longrightarrow$

1
2
3
4
5
.
.
.
.
.
.
.
.
.
.
.
.
.
.
62
63
64

Fig. 3.1

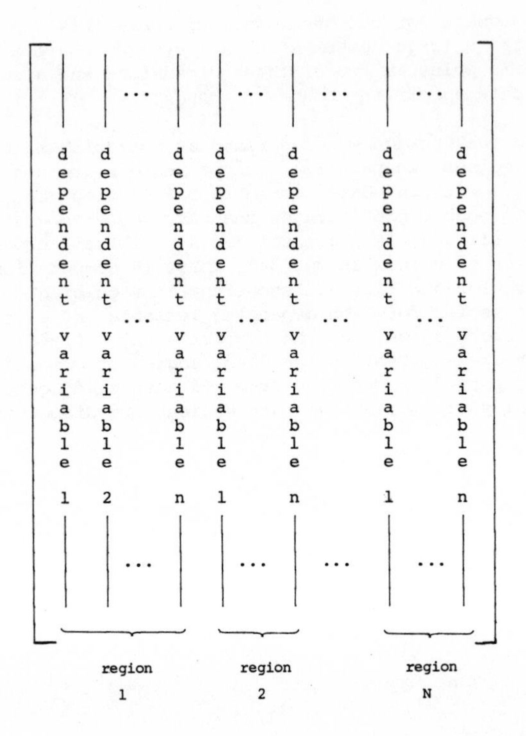

Fig. 3.2

The 'obvious' way to select our regions from a full n×m grid (n and m both multiples of 8) is to take adjacent squares of 64 (8×8) points. This was indeed chosen in the first version of the package. But experience showed that this was inefficient to handle on the DAP because of the problem, noted above, of adjacent points being 'uncoupled' in store. The solution was to further 'uncouple' the regions by taking the 64 points dispersed evenly throughout the computational space. This 'crinkling' of the data, shown in Fig. 4, has the result that, for example, a 4 point difference scheme, with a minimum of 3 by 3 regions (i.e. 9 regions, each of 64 points) can be effected by simple vector arithmetic operations and shifts.

Since the package deals in full with data storage, input and output, this method of storage is not a burden to the user.

5.2.2 Programme Development

The major obstacle in large program verification and development on the DAP is the speed, or rather lack of it, of the DAP assembler. The simplest approach to outputting diagnostic information from the DAP is the DAP Fortran TRACE statement. This has, naturally, to be compiled in order to be executed. Of the three stages in DAP compilation - compile, assemble,

```
10 11 12 10 11 12 10 11 12 10 11 12 10 11 12 10 11 12 10 11 12 10 11 12
 7  8  9  7  8  9  7  8  9  7  8  9  7  8  9  7  8  9  7  8  9  7  8  9
 4  5  6  4  5  6  4  5  6  4  5  6  4  5  6  4  5  6  4  5  6  4  5  6
 1  2  3  1  2  3  1  2  3  1  2  3  1  2  3  1  2  3  1  2  3  1  2  3
10  .  .  .  .  .  .  .  .  .  .  .  .  .  .  .  .  .  .  .  .  .  . 12
 7  .  .  .  .  .  .  .  .  .  .  .  .  .  .  .  .  .  .  .  .  .  .  9
 4  .  .  .  .  .  .  .  .  .  .  .  .  .  .  .  .  .  .  .  .  .  .  6
 1  .  3  .  .  3  .  .  3  .  .  3  .  .  3  .  .  3  .  .  3  .  .  3
10  .  .  .  .  .  .  .  .  .  .  .  .  .  .  .  .  .  .  .  .  .  . 12
 7  .  .  .  .  .  .  .  .  .  .  .  .  .  .  .  .  .  .  .  .  .  .  9
 4  .  .  .  .  .  .  .  .  .  .  .  .  .  .  .  .  .  .  .  .  .  .  6
 1  .  3  .  .  3  .  .  3  .  .  3  .  .  3  .  .  3  .  .  3  .  .  3
10  .  .  .  .  .  .  .  .  .  .  .  .  .  .  .  .  .  .  .  .  .  . 12
 7  .  .  .  .  .  .  .  .  .  .  .  .  .  .  .  .  .  .  .  .  .  .  9
 4  .  .  .  .  .  .  .  .  .  .  .  .  .  .  .  .  .  .  .  .  .  .  6
 1  .  3  .  .  3  .  .  3  .  .  3  .  .  3  .  .  3  .  .  3  .  .  3
10  .  .  .  .  .  .  .  .  .  .  .  .  .  .  .  .  .  .  .  .  .  . 12
 7  .  .  .  .  .  .  .  .  .  .  .  .  .  .  .  .  .  .  .  .  .  .  9
 4  .  .  .  .  .  .  .  .  .  .  .  .  .  .  .  .  .  .  .  .  .  .  6
 1  .  3  .  .  3  .  .  3  .  .  3  .  .  3  .  .  3  .  .  3  .  .  3
10  .  .  .  .  .  .  .  .  .  .  .  .  .  .  .  .  .  .  .  .  .  . 12
 7  .  .  .  .  .  .  .  .  .  .  .  .  .  .  .  .  .  .  .  .  .  .  9
 4  .  .  .  .  .  .  .  .  .  .  .  .  .  .  .  .  .  .  .  .  .  .  6
 1  .  3  .  .  3  .  .  3  .  .  3  .  .  3  .  .  3  .  .  3  .  .  3
10 11 12 10 11 12 10 11 12 10 11 12 10 11 12 10 11 12 10 11 12 10 11 12
 7  8  9  7  8  9  7  8  9  7  8  9  7  8  9  7  8  9  7  8  9  7  8  9
 4  5  6  4  5  6  4  5  6  4  5  6  4  5  6  4  5  6  4  5  6  4  5  6
 1  2  3  1  2  3  1  2  3  1  2  3  1  2  3  1  2  3  1  2  3  1  2  3
10 11 12 10 11 12 10 11 12 10 11 12 10 11 12 10 11 12 10 11 12 10 11 12
 7  8  9  7  8  9  7  8  9  7  8  9  7  8  9  7  8  9  7  8  9  7  8  9
 4  5  6  4  5  6  4  5  6  4  5  6  4  5  6  4  5  6  4  5  6  4  5  6
 1  2  3  1  2  3  1  2  3  1  2  3  1  2  3  1  2  3  1  2  3  1  2  3
```

Fig. 4

consolidate - the first and last require negligible time and can be executed from the terminal, compared with assembly which, even for a short routine, needs to be performed as a batch job. In a package such as PDEPACP which has a modular design, the tracing of an error through a number of subroutines is a time consuming job and makes the assembly stage a considerable obstacle to efficient program testing and development.

5.3 *Output from the DAP*

It is not possible to get data output directly from the DAP. The user is obliged to route all I.O. through the 2980 host. The types of problem which PDEPACP is intended to solve do not have 'an answer', rather we are interested in following the development of a solution, needing output of large quantities of data at intermediate stages in the processing. Further, as a user package, we need to provide warning and error messages and diagnostic data. Judicious placing of TRACE statements, particularly of the character arrays available in the extended versions of Fortran implemented on the 2980 and DAP can solve the latter problem. The former problem, outputting 'DAPfuls' of information, may well be solved by the Block Transfer System which is currently becoming available on the DAP.

6. CONCLUSIONS

The package, as it is implemented at the moment, is still in the relatively early stages of development. The first set of tests, the logical tests, have been completed satisfactorily and at the time of writing this paper the first numerical test has also been completed and the first full test problem is being implemented.

ACKNOWLEDGEMENTS

The authors would like to thank the members of the CFD research group at Salford University, Dr. Gordon Laws, Peter Caine, Christine Hopkins and Ann Murdoch, for their invaluable support and guidance throughout this work. The financial support of the SERC is also acknowledged.

REFERENCES

Stansfield, E. (1981) An analysis to determine relevant data structures in computational algorithms for fluid dynamics problems, 3rd year project, Department of Mathematics, University of Salford.

Walkden, F., Laws, G.T. and Stansfield, E. (1979) SURDATPAC Program Suite, Parts 1-8, Department of Mathematics, University of Salford, Internal Publication.

THE THREE-DIMENSIONAL SOLUTION OF THE EQUATIONS OF FLOW AND HEAT TRANSFER IN GLASS-MELTING TANK FURNACES: ADAPTING TO THE DAP

A.F. Harding and J.C. Carling

(Department of Ceramics, Glasses and Polymers, The University of Sheffield)

ABSTRACT

As an introduction, a description is given of the behaviour of glass in an industrial glass melting furnace. To illustrate the nature of the computational problem, results are shown of previous computations on serial machines. The equations governing the flow and heat transfer are then given, with particular reference to those features which are important in formulating the problem on the DAP. Results of DAP computations on a simplified test problem, involving a three-dimensional rectangular cell, are used to indicate an appropriate numerical solution method. An explanation follows of the way in which the more complex geometry of a throated glass melting furnace has been mapped on to the DAP. Attention is drawn to the many aspects of parallelism which exist and which must be recognized when formulating computational strategy if the DAP is to be used to its full potential. Some examples of the evaluation of (mathematical) vector operators are used to illustrate the concise nature of DAP FORTRAN and the way it differs from conventional code on serial machines. Finally current results of computations of throated furnaces are shown.

1. INTRODUCTION

In industrial production, glass is melted in a tank furnace. The furnace is constructed of refractory material and it is common for it to consist of two cavities joined together by a throat (submerged passage). A typical throated furnace is shown in Fig. 1. The internal dimensions of the two cavities, known as the melting end and the working end, are approximately 6m by 4m and 2m by 4m respectively; the glass depth is typically 1m. Current design practice favours rectangular rather than semicircular working ends. Glass batch (raw material) enters at one end and is melted, usually by gas flames directed over the glass surface. The molten glass, which is at temperatures of the order of 1500°C, then passes through the two cavities by a combination of buoyancy driven circulation and pull (imposed mass flow rate). The production process is continuous. An artist's impression of the behaviour of the glass is shown in Fig. 2. The mathematical models attempt to quantify this picture by solving, in three-dimensions (and in time), the coupled equations of flow and heat transfer. The purpose in developing the models is to improve furnace operation and to assist in the design of future installations.

To illustrate the nature of the buoyancy driven flow, the following results, from a previous computation on a CDC 7600, show how the glass behaves in a typical working end. Fig. 3 shows the idealised

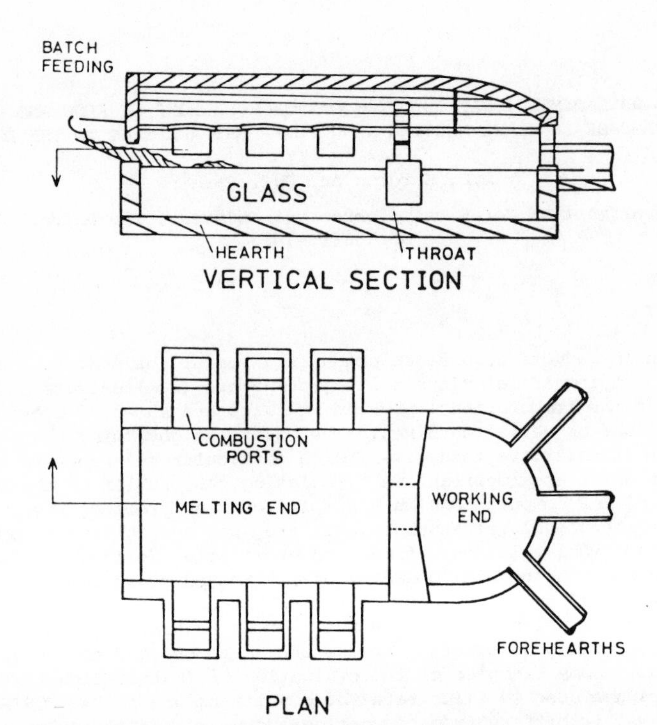

Fig. 1 Cross-fired glass melting tank furnace (From Henderson (1978))

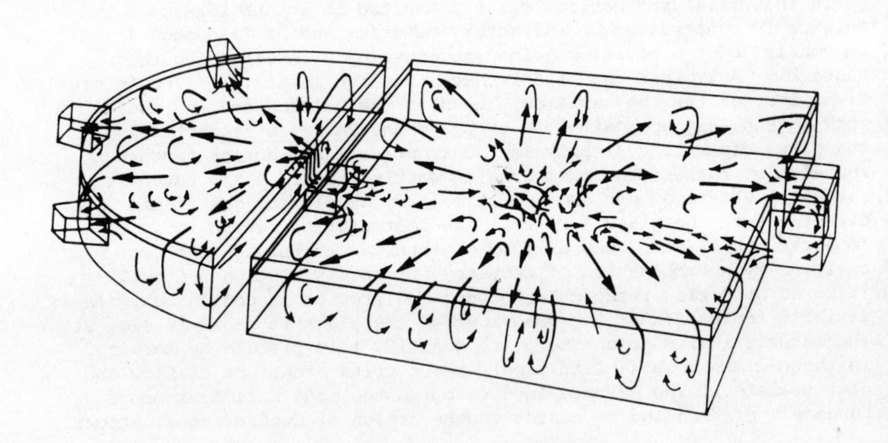

Fig. 2 An artist's impression of the flow in a tank furnace (From Trier (1960))

computational problem. Fig. 4 shows two particle traces, indicating two types of path through this working end, one (no. 1) direct and relatively fast and the other (no. 2) tortuous and slow. The residence times for the two traces are given. This computation will be compared with a DAP computation in section 7.

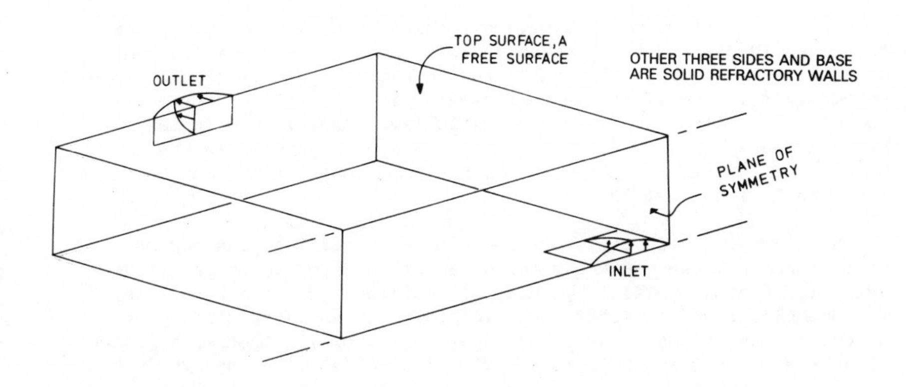

Fig. 3 The computational problem

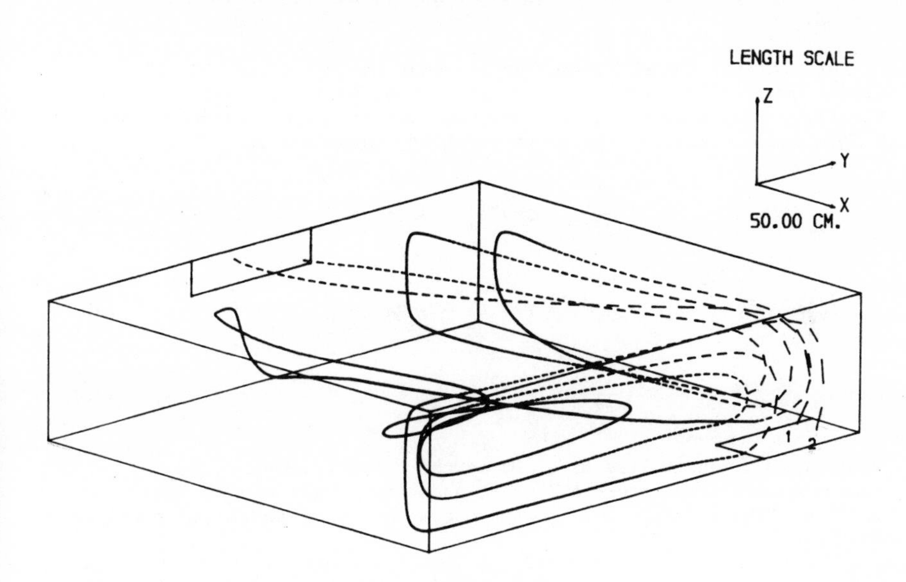

Fig. 4 Particle traces, showing residence times.
Trace no. 1, 26 mins; trace no. 2, 1561 mins.
Local velocity is proportional to the dash length

2. PHYSICAL ASSUMPTIONS AND GOVERNING EQUATIONS

The physical assumptions in the mathematical model are that the flow is Newtonian, incompressible and slow with a Reynolds number of the order of 0.1. The viscosity is temperature dependent and can vary by a factor of 20. Buoyancy driven motion is allowed for by the Boussinesq approximation and the Rayleigh number, Ra, is in the range 10^4 to 10^6. In the energy equation, radiation heat transfer within the glass is dealt with by assuming (with reasonable justification) that the glass is optically thick and then making use of an effective conductivity. The effective conductivity is temperature dependent. Convection is important as well as effective conduction with maximum dimensionless velocity (equivalent to the Peclet number) in the range 10 to 1000. Further details surrounding these assumptions may be found in Carling (1982).

The energy, continuity and three momentum equations can now be established. These equations could be solved directly by primitive variable methods typified by those of Spalding and co-workers (see, for example, Spalding (1981) in which there is an interesting general discussion on the application of numerical models). Another approach, pioneered by Aziz and Hellums (1967), involves eliminating pressure from the momentum equations by introducing a vector vorticity, $\underset{\sim}{\zeta}$, and a vector potential, $\underset{\sim}{A}$. In addition, for open systems, an equation in the scalar potential, ϕ, is required. This method is akin to the stream function and vorticity method familiar in two-dimensions. By historical accident, the method used here is the vector vorticity method.

For clarity, the slow, isoviscous equations are given. In the actual computation, variable viscosity terms have been included in the vorticity equations but these have been omitted here. Without giving a detailed derivation, the three vorticity equations are

$$\frac{\partial\zeta_1}{\partial\tau} = \Pr\nabla^2\zeta_1 + \Pr\mathrm{Ra}\,\frac{\partial\theta}{\partial Y}\,, \tag{2.1}$$

$$\frac{\partial\zeta_2}{\partial\tau} = \Pr\nabla^2\zeta_2 - \Pr\mathrm{Ra}\,\frac{\partial\theta}{\partial X}\,, \tag{2.2}$$

and

$$\frac{\partial\zeta_3}{\partial\tau} = \Pr\nabla^2\zeta_3, \tag{2.3}$$

where τ is dimensionless time, Pr is the Prandtl number, θ is the dimensionless temperature and other quantities have already been defined. The equation for the three components of vector potential is

$$\nabla^2\underset{\sim}{A} = -\,\underset{\sim}{\zeta}. \tag{2.4}$$

The equation for the scalar potential is

$$\nabla^2\phi = 0. \tag{2.5}$$

The energy equation is

$$\frac{\partial\theta}{\partial\tau} = \underset{\sim}{\nabla}.(\kappa\underset{\sim}{\nabla}\theta - \underset{\sim}{U}\theta), \tag{2.6}$$

where κ is the dimensionless effective conductivity and $\underset{\sim}{U}$ the velocity vector. By definition the following relationships connect $\underset{\sim}{\zeta}$, $\underset{\sim}{A}$, ϕ and $\underset{\sim}{U}$:

$$\underset{\sim}{\zeta} = \underset{\sim}{\nabla} \wedge \underset{\sim}{U}, \tag{2.7}$$

and

$$\underset{\sim}{U} = \underset{\sim}{\nabla} \wedge \underset{\sim}{A} - \underset{\sim}{\nabla}\phi \tag{2.8}$$

Boundary conditions are not described here. It is not appropriate to give such detail since the main purpose is to discuss overall computational strategy in adapting to a new machine (the DAP).

When approaching problems on the DAP, it is vital at least to recognise the many aspects of parallelism which exist in almost any computation. Whether a particular parallel feature is then used becomes a matter of computational strategy. Thus it should be pointed out that in equations (2.1) to (2.3) the three components of vorticity are mutually independent. These three equations could, if required, be solved in parallel. Similarly each component equation connecting $\underset{\sim}{\zeta}$ and $\underset{\sim}{A}$ (equation (2.4)) could be treated in parallel. This will be referred to in more detail shortly.

It should be noticed in passing that on serial machines the vector vorticity method is arguably disadvantageous since seven flow equations (in $\underset{\sim}{\zeta}$, $\underset{\sim}{A}$ and ϕ) have to be solved as opposed to four (in $\underset{\sim}{U}$ and pressure) with the primitive variable method. Noting that the scalar potential, ϕ, is often removed from the computation (it is zero for closed systems and often read in as data), it follows that if components of vectors are treated in parallel on the DAP then both methods reduce to effectively two equations. It would be interesting to investigate this further but, whatever the argument, the vector vorticity method has been used here.

The governing equations (2.1) to (2.6) are solved by finite difference methods on a uniform, rectangular grid. In each cell within the grid, the variables $(\zeta_1,\zeta_2,\zeta_3)$, (A_1,A_2,A_3), ϕ and θ are located as shown in Fig. 5. This is the geometric storage system. The reason for this elaborate staggered system is that it enables velocities to be calculated naturally from equation (2.8) using straightforward differences. The components of $\underset{\sim}{A}$ and ϕ are stored precisely where they are required to form the curl and the grad operators. There are other advantages in using this system, particularly concerning the application of boundary conditions, but these will not be discussed.

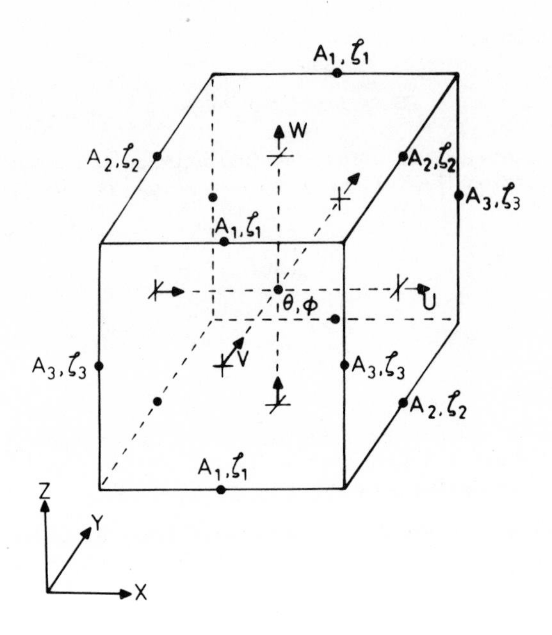

Fig. 5 The geometric location of the dependent variables

3. GENERAL COMPUTATIONAL STRATEGY AND CHOICE OF NUMERICAL METHOD

The general computational strategy adopted on the DAP is to regard each X-Y plane of each dependent variable as being stored in parallel on DAP planes (i.e. as a DAP matrix). The storage in the Z direction is built up by layers of these planes one above the other as shown in Fig. 6. The question of treating the three components of vectors in parallel will be left aside for the moment.

A suitable numerical solution method must now be chosen. Since it is not certain that methods which perform well on serial machines will transfer readily to the DAP, it is essential to reassess the commonly used numerical schemes without preconceptions. So, to investigate this a preliminary test problem has been devised. The following Poisson equation has been chosen as representative of most of the equations given in section 2:

$$(\partial\theta/\partial\tau=)\ \nabla^2\theta+1 = 0. \tag{3.1}$$

Boundary conditions are $\theta = 0$ on all boundaries except K=KMAX (see Fig. 6) where $\partial\theta/\partial Z = 0$ is used. A 64 by 64 by 10 grid has been specified with grid spacing DX=DY=0.129 and DZ=0.125. This gives a length to depth ratio typical of the overall furnace dimensions shown in Fig. 1. The DAP FORTRAN variable THETA(,,K) has been used, indicating the parallelism. It is worth describing the numerical methods in more detail, particularly their implementation on the DAP.

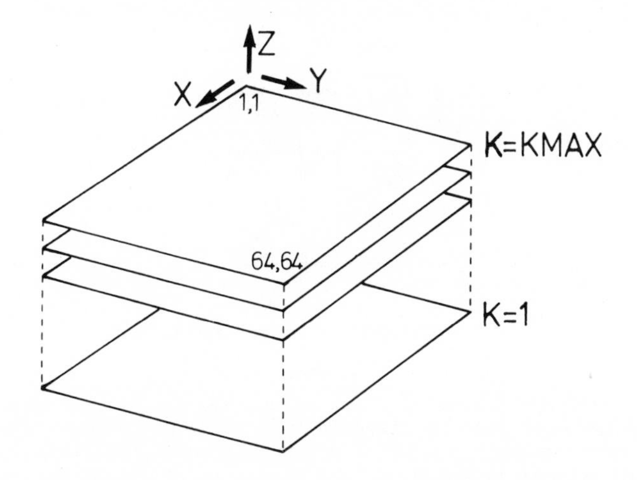

Fig. 6 General computational storage scheme

3.1 Numerical methods

Three numerical methods have been used to solve the test problem: an explicit time-dependent method which solves equation (3.1) in the asymptotic limit as time becomes large; an alternating direction implicit (ADI) method again applied in the asymptotic limit but with a larger time step; and an iterative method, successive over relaxation (SOR) with chessboard ordering which solves the steady state problem directly. All three methods have been used to solve the test problem on the DAP and for comparison some of the methods have been run on a CRAY 1, a CDC 7600 and a PRIME 750. A paper by Webb (1980) describes a similar application of these methods in two dimensions.

The explicit method is trivial to program on any machine but it is particularly suited to the DAP. The expression for new values of θ is given in terms of old values at surrounding grid points thus enabling the calculation on each X-Y plane to be carried out in parallel.

The ADI method involves the inversion of tridiagonal matrices formed by treating the solution implicitly in each coordinate direction in turn. Intuitively this method seems unsuited to the DAP. Nevertheless routines exist to solve tridiagonal systems and the DAP library routine FO4ITTRIDS64 has been used in the X and Y directions with conventional Gaussian elimination in the Z direction.

The SOR method with chessboard ordering is well known to historians of finite difference techniques (it is referred to in Smith (1965), for example) but it is probably true to say that until recently there has been no need to use it; the conventional ordering scheme being most appropriate on serial machines. The idea (in two-dimensions first) is to label squares alternately black and white, like a chessboard, as shown in Fig. 7. If the chessboard is regarded as a DAP plane then the calculation can be performed in parallel as

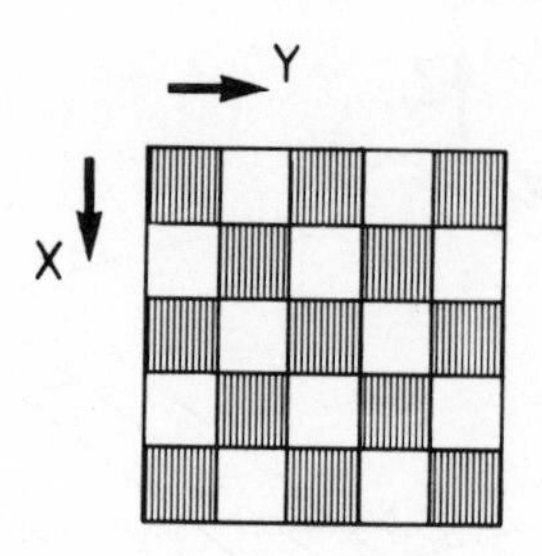

Fig. 7 Chessboard ordering in two-dimensions

follows: first new values of θ are calculated for all the white squares in parallel using a standard five point SOR formula which requires only those old values of θ on neighbouring black squares; then the process is repeated and new values on all the black squares are calculated using the values on the white squares just obtained. The degree of parallelism in this two stage operation is only one half that of the explicit method since only half the new values can be worked out at a time. The advantage of the SOR method is that fewer iterations are required.

Moving on to the three-dimensional case, there are two ways of extending the idea of chessboard ordering. Fig. 8(a) shows a fully three-dimensional chessboard and Fig. 8(b) a stack of two-dimensional chessboards in which there are entirely black and entirely white columns in the Z direction. Both these arrangements have been used here. The DAP version of SOR already described in two-dimensions extends naturally to the fully three-dimensional chessboard. In the case of the stack of two-dimensional chessboards, the SOR method proceeds on each X-Y plane as before, but uses recently calculated values on the plane beneath (assuming the advance is in the positive Z direction) as soon as they become available. The degree of parallelism in both these approaches is again one-half that of the explicit method. However, the degree of parallelism in the fully three-dimensional chessboard can be restored by noticing that adjacent X-Y layers of black cubes (or of white) can be processed in parallel. This is achieved by swapping the DAP storage locations corresponding to all the white cubes on a given plane with all the black cubes on the plane directly above. This affects the DAP storage only; it does not alter the way in which the SOR method operates. The swapping technique, pointed out by Reddaway (1982), has not been used here but it is under investigation.

A further extension to SOR has also been investigated: line SOR (or LSOR) with chessboard ordering in the X-Y plane and the lines (corresponding to the stacks) running in the Z-direction. Gaussian elimination can then be used to solve for all points along a line simultaneously.

Hockney (1982) pointed out that equations like (2.4) could be solved rapidly in rectangular regions by fast Fourier transform methods. The solutions in sub-regions (the melting end, throat and

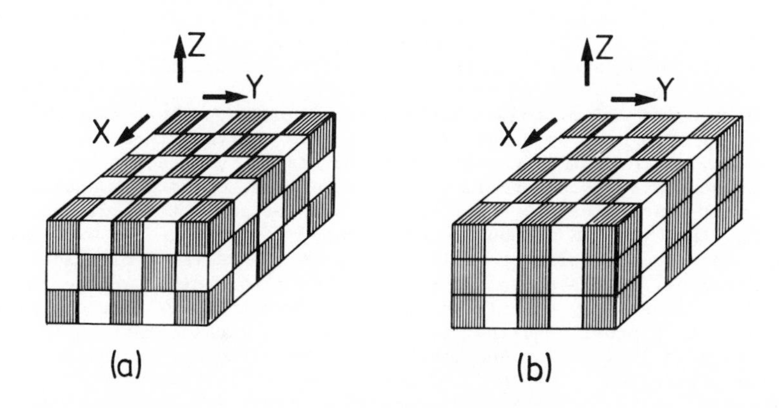

Fig. 8 Chessboard ordering in three-dimensions. (a) Fully 3-D chessboard. (b) A stack of 2-D chessboards

working end, in this particular furnace) could then be interconnected within an iterative scheme. Whether this method would be satisfactory for the present purpose remains unknown.

3.2 Test results

Table 1

Execution times (seconds) on various machines for various methods. In all cases a convergence criterion of 5×10^{-4} has been used.

Method	No. of iterations to convergence	DAP time per iteration	DAP total execution time	CRAY 1 total time	CDC7600 total time	PRIME 750 total time
Explicit	820	0.019	15	7	65	1640
ADI	83	0.350	29	-	69	1360
SOR (opt.)	31	0.033	1	-	-	-
LSOR (opt.)	31	0.081	2.5	-	-	-

Details of the results are given in table 1. The first point to make is that there is unquestionably more work to be done in the investigation of numerical methods. There are a number of omissions in the table and it is not entirely clear whether sensible comparisons are being made. These are the proceedings of a workshop, however, and displaying work in progress should therefore be regarded as a praiseworthy activity! Looking first at the explicit method and comparing the execution times on the various machines, the DAP is 110 times faster than a PRIME 750, 4.3 times faster than a CDC 7600 and slower, by a factor of 2, than a CRAY 1. On this particular test problem the DAP architecture is being fully exploited, so too is the CRAY's, and the comparison is perhaps unduly severe on the two serial machines.

Comparing the ADI method with the explicit method, shows that on the DAP ADI is slower by a factor of 2, on the CDC the times are comparable and on the PRIME the ADI method is faster by a factor of about 1.2. These figures are revealing since experience on serial machines in the past has indicated, admittedly on smaller scale three-dimensional problems, that the ADI method is considerably faster than the explicit method. An advantage factor as low as 1.2 on the PRIME, a classical serial machine, is therefore surprising. The slight disadvantage on the CDC can presumably be attributed to the fact that ADI disrupts the small amount of pipelining that takes place on a machine that in most other respects is serial. The disadvantage of using ADI on the DAP is clearly revealed. The DAP does not digest tridiagonal matrices particularly well (in the X and Y directions, that is), taking 18 times longer per time step than the explicit method. The considerable effort involved in programming ADI on the DAP would appear to be wasted.

The final comparison that emerges from table 1 is the extremely satisfactory performance of the point SOR method. The figures refer to an optimised SOR method with chessboard ordering of the stacked variety. The execution time for a fully three-dimensional chessboard was slightly slower. However, when the two arrangements were applied to a different problem involving derivative conditions at all boundaries, the fully three-dimensional chessboard method was far superior, the stacked chessboard method requiring approximately ten times the number of iterations. Presumably extending the parallelism still further by swapping planes in the way described in section 3.1 would again improve the speed of computation. The balance of the argument therefore favours the fully three-dimensional arrangement.

The table also shows results using LSOR. It can be seen that the number of iterations is the same as point SOR while the time per iteration and the total time has increased. This is because the rate of convergence is still restricted by the rate at which the solution is propagated across the X-Y plane. On the DAP LSOR is likely to be superior to point SOR only if there are more mesh points in the Z direction than in either the X- or Y-directions.

Commenting on the test problem itself, it should be said that it would be desirable to repeat the series of runs on a problem containing a first order term in the governing equation. In such circumstances, ADI might be expected to perform better than in the present example. This work is in progress.

4. MAPPING A THROATED FURNACE ON TO THE DAP

Returning to the furnace problem, Fig. 9 shows the idealised throated furnace which is to be computed on the DAP. The general ideas on DAP storage remain as described in section 3, but with one modification. Since the length of the furnace is approximately three times the width and remembering the discussion in section 2 on the parallelism within the vector equations (see equations (2.1) to (2.4)), it is desirable to divide the DAP storage area into three parts.

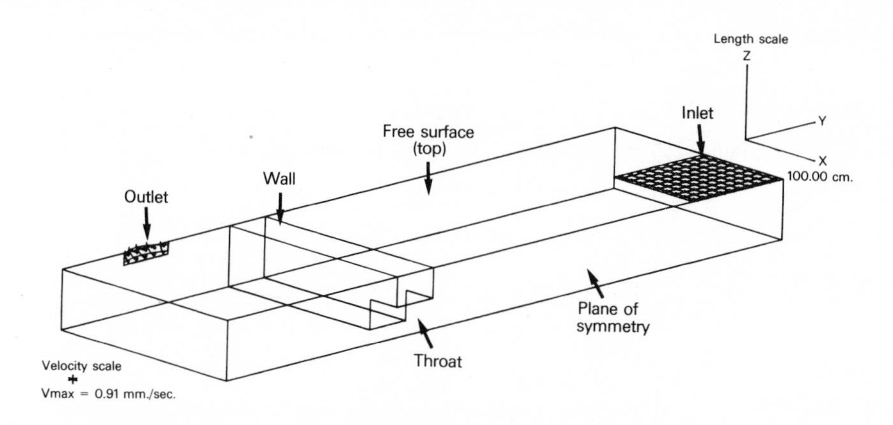

Fig. 9 The idealised throated furnace

This division is shown in Fig. 10. Vector quantities $\underset{\sim}{\zeta}$ and $\underset{\sim}{A}$ are stored so that, for example, ζ_1 is stored between X = 1 and 21, ζ_2 between 22 and 42 and ζ_3 between 43 and 63. X = 64 is redundant. A 64 by 63 DAP would be more suitable for this application! Scalar quantities, like θ and ϕ, are stored three times (triplicated), once in each region. Operations on the mathematical scalars by DAP scalars are performed in all three regions since, per DAP plane, it takes no longer than in one region only. The triplicated storage is advantageous when mathematical scalars are required in vector equations.

The geometry of the wall and the throat is dealt with by using masks (logical matrices), for example, INSIDE(,,K) which is .TRUE. within the glass but .FALSE. round the boundaries and within the wall. As K varies, INSIDE(,,K) adjusts to accommodate the geometry of the wall and throat. The masks (other masks are determined from INSIDE(,,K)) can be used to ensure that particular equations are solved in appropriate places, that evaluations are restricted to particular regions and that boundary conditions are properly applied. This method of mapping the internal furnace geometry is extremely powerful and easy to implement. Looking forward to further applications, a furnace design often includes more than one wall, subterranean throats, steps in the base and electrodes protruding into the chambers (to provide Joule heating). All these features can be implemented on the DAP very easily by changing INSIDE(,,K), from which other masks are automatically derived. To do this on a serial or vector machine would involve either lengthy program construction or the use of logical arrays and inefficient conditional statements.

5. EVALUATING DIFFERENTIAL OPERATORS IN DAP FORTRAN

With the storage system described in section 4, a number of DAP FORTRAN programming subtleties are involved in evaluating the differential operators. The grad, div, Laplacian and curl operators are considered in turn. A paper by Wilson (1982) is relevant in this context.

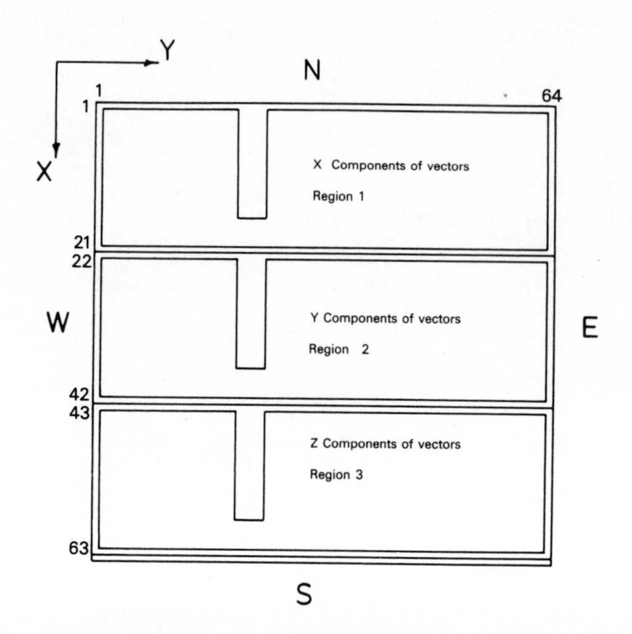

Fig. 10 Storage arrangement within the DAP

5.1 *The grad operator*

As an example, consider $\nabla\phi$ which occurs in equation (2.8). To evaluate each velocity component (leaving aside for the moment the contribution of $\nabla \wedge \mathbf{A}$) requires the following operations in each region:

in region 1, X component of $\mathbf{U}$: -(PHI(+,,K)-PHI(,,K))/DX ;
in region 2, Y component of $\mathbf{U}$: -(PHI(,+,K)-PHI(,,K))/DY ;
in region 3, Z component of $\mathbf{U}$: -(PHI(,,K+1)-PHI(,,K))/DZ.

Note that this is why scalar variables, like ϕ, are required in all three regions. In DAP FORTRAN $\nabla\phi$ can be evaluated in parallel using the MERGE function [1]. A single statement is all that is required to calculate the vector velocity on each DAP plane.

```
  U(INSIDE(,,K),K)= -(MERGE(PHI(+,,K),
 -                    MERGE(PHI(,+,K),PHI(,,K+1),REGION2),
 -                    REGION1)
 -                -PHI(,,K))
 -                /MERGE(DX,MERGE(DY,DZ,REGION2),REGION1)
```

where REGION1 and REGION2 are logical matrices which define the rectangular regions between X=1 and 21 and X=22 and 42 respectively.

[1] The DAP FORTRAN function MERGE operates as follows: if A=MERGE(B,C,LM) then matrix A:=matrix B where logical matrix LM is .TRUE. and A:=matrix C elsewhere.

This statement would occur within a DO loop in K. The presence of INSIDE(,,K) restricts the computation to locations within the glass only.

5.2 The div operator

Consider the term $\underset{\sim}{\nabla}.(\kappa\underset{\sim}{\nabla}\theta)$ which occurs in the energy equation (2.6). Put $\kappa\underset{\sim}{\nabla}\theta \equiv (Q_1,Q_2,Q_3)$, assume that $\underset{\sim}{Q}$ has been evaluated as in section 5.1 and that Q123(,) exists with Q_1, Q_2, and Q_3 stored in each region. To calculate $\underset{\sim}{\nabla}.\underset{\sim}{Q}$, first perform the operation defined in section 5.1 as if Q123(,) were a mathematical scalar. The result is a quantity, DQDXYZ(,), with $\partial Q_1/\partial X$, $\partial Q_2/\partial Y$ and $\partial Q_3/\partial Z$ stored in each region. To obtain $\underset{\sim}{\nabla}.\underset{\sim}{Q}$, a mathematical scalar, the three components must be added together and triplicated, like any other scalar, being stored once in each region. On a 64 by 63 DAP, the right hand side of a statement to do this would be as follows:

```
DQDXYZ + SHNC(DQDXYZ,21)+SHSC(DQDXYZ,21),                    (5.1)
```

where SHNC and SHSC are north circular and south circular shift functions respectively. On a 64 by 64 DAP extra shifts have to be inserted. Although this appears straightforward, the shift function is expensive in bit operations. This will be discussed further in section 5.5.

5.3 The Laplacian

The Laplacian ($\nabla^2\theta$, for example) is merely a special case of the operator described in section 5.2. It could therefore be evaluated in exactly the same way but, because of the difficulty with the shifting operation, it is in fact more efficient simply to triplicate the direct application ∇^2 without resorting to the vector components involved in $\underset{\sim}{\nabla}.\underset{\sim}{\nabla}\theta$. This might not be the case, however, if an alternative storage system is used (see section 5.5).

When expressions like $\nabla^2\underset{\sim}{A}$ are being evaluated there is no choice, of course, so the direct application of ∇^2 must be used. The operation is performed on each component of $\underset{\sim}{A}$ in parallel in the three regions.

5.4 The curl operator

In outline, the operation $\underset{\sim}{\nabla} \wedge \underset{\sim}{A}$, or

$$\frac{\partial A_3}{\partial Y} - \frac{\partial A_2}{\partial Z}, \frac{\partial A_1}{\partial Z} - \frac{\partial A_3}{\partial X}, \frac{\partial A_2}{\partial X} - \frac{\partial A_1}{\partial Y},$$

requires: A_3, A_2 in region 1; A_1, A_3 in region 2; and A_2, A_1 in region 3. This is achieved by intermediate matrices

```
        ASHN(,,K)=SHNC(A(,,K),21)
                                              (5.2)
and     ASHS(,,K)=SHSC(A(,,K),21),
```

again assuming a 64 by 63 DAP. ASHN now contains A_2, A_3 and A_1 in regions 1, 2 and 3 respectively, whereas ASHS contains A_3, A_1 and A_2. Using the MERGE function (as in section 5.1) to evaluate $\underset{\sim}{U} = \underset{\sim}{\nabla} \wedge \underset{\sim}{A}$ gives

```
 U(INSIDE(,,K),K)=(MERGE(ASHS(,+,K),
-                        MERGE(ASHS(,,K+1),ASHS(+,,K),REGION2),
-                        REGION1)
-                       -ASHS(,,K))
-                       /MERGE(DY,MERGE(DZ,DX,REGION2),REGION1)
-                 -(MERGE(ASHN(,,K+1),
-                         MERGE(ASHN(+,,K),ASHN(,+,K),REGION2),
-                         REGION1)
-                        -ASHN(,,K))
-                        /MERGE(DZ,MERGE(DX,DY,REGION2),REGION1)
```

5.5 Remarks on the evaluation of vector operators

In section 5.2 it was mentioned that shifts over long distances are computationally expensive. The amount of shifting can be reduced if the storage of the three regions is 'crinkled', so that corresponding components of vector quantities are stored at adjacent DAP locations (Parkinson (1982)). This means that the shifts of 21 in the code for finding div and curl (see statements (5.1) and (5.2)) can be replaced by shifts of one or two, but that shifts of 3 would be required for finding X-derivatives. Since the latter occur much more often, it is not clear whether crinkled storage is an advantage, particularly as program clarity would be affected. This is under investigation.

It should also be pointed out that the use of this crinkled storage system would almost certainly change the finding in section 5.3 on the most efficient way to evaluate $\nabla^2\theta$. The use of vector components in formulating $\underset{\sim}{\nabla}.\underset{\sim}{\nabla}\theta$ would probably be beneficial when used in conjunction with crinkling. Whether this method (which carries a penalty in additional X shifts) would be faster than the storage system used here and the direct application of ∇^2 (which is triplicated as described in section 5.2) remains the subject of experiment.

It should be remembered that the difficulty with shifting arose in the first place only because of the existence of the three regions. This is itself a matter of choice. It could be that storing components of vectors as separate DAP matrices would be a better strategy. Although this would seem unlikely in the present problem with elongated geometry, such an arrangement would have to be considered if the plan dimensions were approximately equal and it should not be ruled out.

6. RESULTS

The flow and temperature distribution in the throated furnace (see Fig. 9) have been computed using SOR with fully three-dimensional chessboard ordering, the storage arrangements detailed in section 4 and the computational techniques given in sections 3 and 5. Brief details of the physical parameters and boundary conditions on temperature are as follows. The value of Ra is 7000; the pull is 0.28 kg/s; U_{max}, which occurs at the outlet, is 60; the glass temperature at the

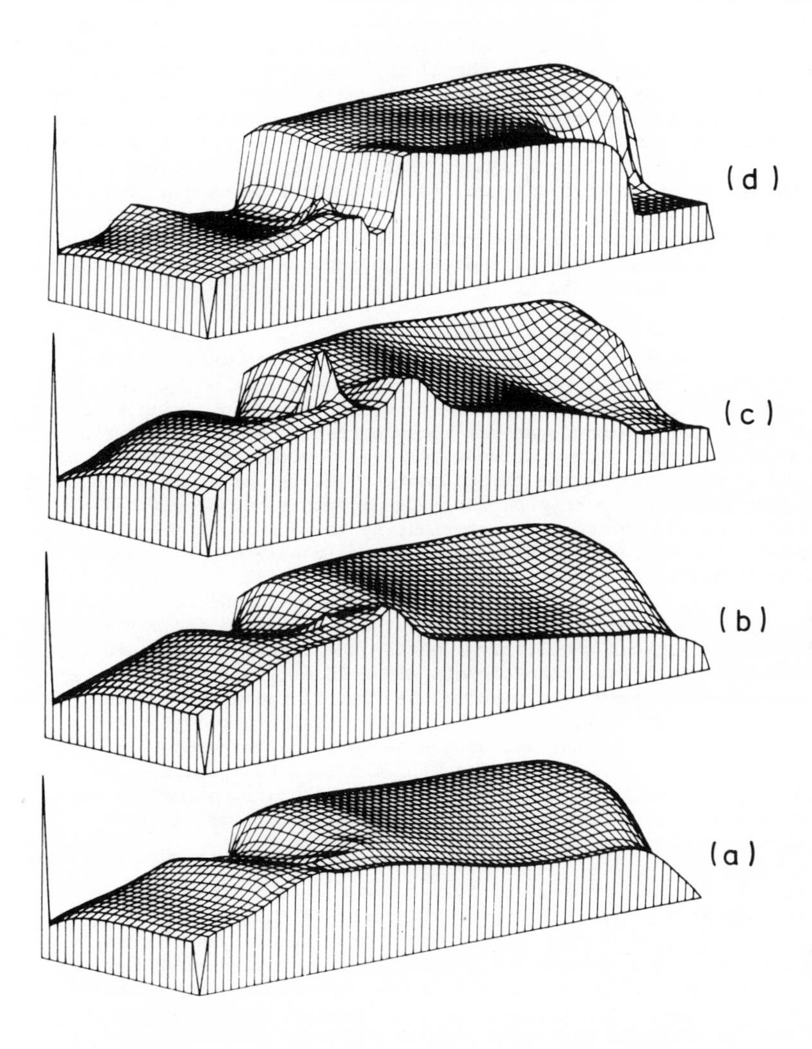

Fig. 11 Isoplots of temperature. (a) On K=1, near base. (b) On K=5, near the roof of the throat. (c) On K=8, part way between glass surface and throat roof. (d) On K=11, near glass surface. The spike to the left of each isoplot gives the temperature scale which is, in each case, 900°C at the base and 1600°C at the tip

inlet is fixed at 1000°C; the gas temperature in the space above the melting end is 1500°C; and the gas temperature in the space above the working end is 1100°C. Heat exchange takes place between each gas space and the free surface of the glass; heat losses through the refractory side walls and the base are used to determine conditions on $\partial\theta/\partial n$ at glass boundaries in contact with refractories. It should be pointed out that this set of data constitutes a preliminary attempt at modelling an operating furnace. For example, the Rayleigh number is perhaps an order of magnitude less than might be encountered in practice. In addition, there is some doubt as to whether the free surface temperature conditions realistically reflect the behaviour of a full scale plant. Work is underway to improve the degree of realism in the mathematical model.

Fig. 11 shows isoplots of temperature on various planes. In part (a) the influence of the heat loss through the side walls and the dividing wall can be seen. In part (b) a local peak within the throat indicates the presence of relatively strong convection bringing glass from the hotter regions near the glass surface. Part (c) shows a numerically stable, but spurious, spike just within the working end and above the level of the throat roof. This is not to be confused with the spike that occurs on the left of each isoplot, which is simply a scaling device. The spurious spike definitely comes into the category of work in progress! It is probably caused by an excessive velocity (and hence excessive convection) resulting from errors in the prediction of the corner and edge vorticity in the re-entrant geometry encountered on emerging from the throat into the working end. This numerical peculiarity, which others have experienced in two-dimensional flows (see Holstein and Paddon (1981), for example) is being investigated. In part (d) of Fig. 11 the effect of the glass surface boundary conditions can be seen. The influence of the fixed inlet temperature and the difference between the temperatures in each gas space can be seen. Other features are the steep temperature gradient through the dividing wall, the (less spurious) convection just above the level of the throat roof and the effect (another local peak) of the glass flowing through the exit.

Finally Fig. 12 shows velocity vectors near the glass surface and on the symmetry plane. The symbols indicate a change in plotting scale as indicated in the legend. Note that the buoyancy driven circulation in the working end (in Fig. 12) is less marked than that in Fig. 4. This is a consequence of the reduced Rayleigh number (7000) in the DAP computation; the previous computation on the CDC 7600 having been performed with Ra = 20,000.

The computation took approximately 500 secs of DAP time involving 740 iterations of the SOR method. A convergence criterion of $\Delta q/(q_{max}-q_{min})$ everywhere less than 5×10^{-5} was used, where Δq is the change per iteration in each dependent variable q.

7. CONCLUSIONS

The DAP is ideally suited to glass furnace computations and to three-dimensional flow and heat transfer in general. Programs of great power and versatility can be written provided a radical approach is adopted. The most important step taken in this investigation was

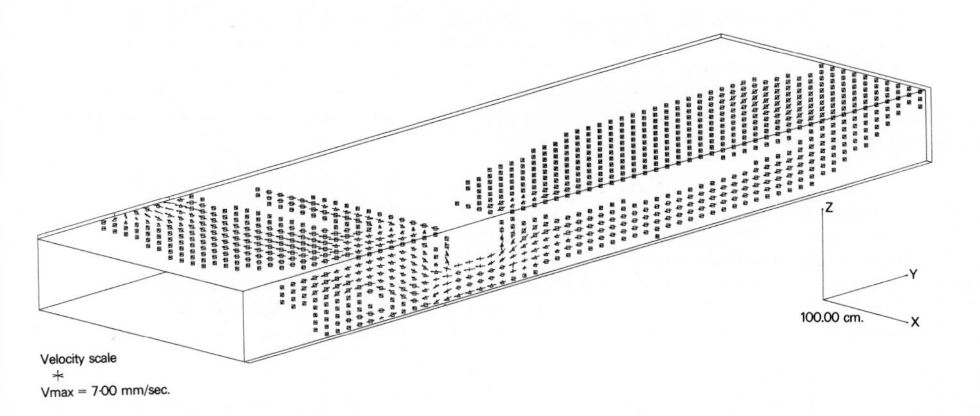

Fig. 12 Velocity vectors near the glass surface and on the symmetry plane. The symbols indicate the plotting scale as follows: O, as given; +, the length of vectors is scaled up to twice the given scale; Δ, four times; □, eight times.

to discard all previous programs along with all preconceptions. Such drastic measures must be taken if the DAP is to be used to its full potential. If an attempt is made merely to adapt existing programs then the performance of the DAP will probably fall below expectations.

It is not possible to make more than loose comparisons between the present furnace computations on the DAP and those on other machines. For example, when comparing the throated furnace computation on the DAP with the working end computation on the CDC 7600 described in section 1, first there is a factor of approximately four in the size of the computation. One problem (on the DAP) used SOR and the other ADI; one contained geometric complications (a wall and throat) and the other did not; and, in addition, the Rayleigh numbers for the two computations were not the same. It is interesting, nonetheless, to note that the computing times on the two machines were roughly equal. This evidence suggests (loosely) that the DAP is approximately four times faster than a CDC 7600 on this type of computation. This agrees roughly with the figure of 4.3 for the test problem in section 3.2 (see table 1), although the performance of the serial machine would almost certainly deteriorate when dealing with the more complicated geometry, but by how much remains unclear.

There is not sufficient evidence in the present paper to compare the performance of the DAP with the CRAY 1, since the only results obtained on the CRAY involved a simplified test problem. Nevertheless, a few comments can be made on the way in which the CRAY might adapt to furnace modelling. The introduction of the more complicated geometry would adversely affect the CRAY's vector processing capacity. The furnace program uses upwind differences which deal with the first order convection terms in the energy equation. These were not present in the test problem, in which a Poisson equation was used. The DAP deals with upwind terms straightforwardly by using a single shift operation where necessary. The CRAY would either have to use a conditional statement within nested DO loops or avoid such heresy by performing more arithmetic. No further attempt is made here to assess the balance of power.

In any case making such comparisons is perhaps missing the point. There are questions to do with setting up the problem and program structure both of which affect program development time. In problems involving three-dimensional geometry, the use of masking is such a natural and direct approach that this alone would seem good reason to favour the use of the DAP. Treating components of mathematical vectors in parallel, just as they are in vector analysis, also seems sensible.

The drawback (or advantage?) of the DAP is that it is a special purpose machine. A special purpose dialect of FORTRAN has to be learnt and new programs developed. Without question is that the capital cost of the DAP is at least an order of magnitude less than other super computers. A workstation with a DAP attached would appear to offer an extremely powerful computing tool: a network of 64 by 63 DAPs all doing glass furnace computations!

ACKNOWLEDGEMENTS

Comments received at the workshop have been incorporated in the text. The authors would like to thank R. Hockney, H. Holstein, J. Marsh, D. Paddon, D. Parkinson, S. Reddaway and A. Wilson for their suggestions. Financial support from a SERC Research Grant and from all the major UK glass manufacturing companies is gratefully acknowledged.

REFERENCES

Aziz, K. and Hellums, J.D. (1967) Numerical solution of the three-dimensional equations of motion for laminar natural convection, *Phys. of Fluids*, **10**, pp. 314-324.

Carling, J.C. (1982) A reappraisal of mathematical modelling of flow and heat transfer in glass tank forehearths, *Glass Technol.*, **23**, pp. 201-222.

Henderson, W.D. (1978) Numerical and experimental studies of heat and fluid flow processes in glass melting tank furnaces and fore hearths, Ph.D. Thesis, Department of Ceramics, Glasses and Polymers, University of Sheffield.

Hockney, R. (1982) Private communication at the Workshop.

Holstein Jr., H. and Paddon, D.J. (1981) A singular finite difference treatment of re-entrant corner flow, *J. Non-Newtonian Fluid Mech.*, **8**, pp. 81-93.

Parkinson, D. (1982) Private communication at the Workshop.

Reddaway, S.F. (1982) Private communication at the Workshop.

Smith, G.D. (1965) Numerical solution of partial differential equations, Oxford University Press, p. 150.

Spalding, D.B. (1981) A general purpose computer program for multi-dimensional one- and two-phase flow, from Mathematics and Computers in Simulation XXIII, North-Holland, pp. 267-276.

Trier, W. (1960) Glass currents: their significance and measurement, *Glastechn. Ber.*, **33**, pp. 401-411.

Webb, S.J. (1980) solution of elliptic partial differential equations on the ICL Distributed Array Processor, ICL Technical Journal, pp. 175-190.

Wilson, A. (1982) Examples of array processing in the next FORTRAN, from The relationship between numerical computation and programming languages (J.K. Reid, ed.), North-Holland, pp. 179-183 and these proceedings.

DOCUMENT ABSTRACTING ON THE DISTRIBUTED ARRAY PROCESSOR

D.E. Oldfield

(International Computers Limited and currently at Computing Laboratory, University of Kent at Canterbury)

1. INTRODUCTION

This paper describes part of a research project on various aspects of processing natural language text on the Distributed Array Processor (DAP) (Flanders, Hunt, Reddaway and Parkinson (1977), Parkinson (1980)). It describes a method of abstracting for which a parse is first performed. The syntactic category of each word is encoded into a bit-significant 64-bit code, which can be manipulated by simple logical operations and reduces data movements to logical shifts. This technique is well suited to the DAP which operates very efficiently on bit-level data; and by suitable choice of data mapping, this in turn allows the analysis of a large volume of text at one time.

2. BACKGROUND

Text handling has been a comparatively neglected area of parallel processing, and it is one of the aims of this project to extend the DAP's application repertoire in this direction.

This work is loosely based on some earlier work at the University of Kent at Canterbury (UKC), which was sponsored by the British Library to classify documents such as Patents, British Standards and Statutes, and store them in a database so that the information could be accessed in a consistent manner. In their report on this project, Wilson and Ball (1974) suggested further investigation into, among other things, clause-level syntax analysis, with a strong recommendation to use texts from Statutes in Force. This source is chosen for the following reasons:

(a) if something is referred to more than once, the same word is repeated, rather than using synonyms which is the tendency in informal text; this reduces the need for comparing word meanings via a thesaurus.

(b) the sentences, while often very long and logically complex, are generally well structured and grammatically correct;

(c) the number of different verbs used is small and the main verb in a sentence often has a modal which can easily be detected;

(d) the use of commas is well regulated compared with many texts in which they tend to be used erratically.

Once a parse has been done there are various uses to which it can be put. For example, for document classification/identification an abstract may be taken. (Another example - information retrieval is discussed later.) Such an abstract would be quite unlike those produced by previous automatic abstracting systems.

While automatic abstracting or extracting has been studied since the pioneering work of Luhn (1958), the topic seems to have progressed little over the last ten years. The kinds of abstract that Luhn pointed the way to, and which were developed by Edmundson (1964) and others in the sixties, were based on selecting whole sentences from a text on the basis of "significant content". For example, Edmundson (1969) describes an improved method for selecting significant sentences in a text based on the frequency of a variety of categories of key words. Later studies have incorporated some syntax analysis in order to reject some parts of sentences, combining this with a weighting system to be applied to whole sentences (Mathis, Rush and Young (1973)).

3. PROBLEM DEFINITION

I have set out to take some texts from a well-defined subject-area and abstract from them a set of sentences with the following properties:

(a) Each input sentence should be transformed into a corresponding output sentence (except that there will be a few which the system is unable to parse, and these should be discarded).

(b) Each output sentence should be grammatically correct. It should be the kernel of the related input sentence consisting of the main part of the subject noun phrase and the main verb or verb phrase from that sentence including any significant "object" phrases which may be present, but excluding prepositional qualifiers in noun phrases.

The abstract is intended to be an indication of the document's contents, with every sentence of the original being represented. It is proposed that this abstract be used as a retrieval tool to obtain further information in a full-text environment.

4. METHOD

4.1 Linguistic considerations

The text is input in free format, just as it might be held in a computer filestore ready for transmission to a lineprinter. Also available is a lexicon of function words (e.g. prepositions, determiners, auxiliary verbs, etc.). The category of each function word is held as a 64-bit code associated with its entry in the lexicon. The number of entries in the lexicon is kept to a minimum in order to reduce the time taken to perform the dictionary lookup. There are currently about 180 words in the lexicon, but with further refinement of the system it could grow to about twice this size.

When the text window and the lexicon have been read into the DAP store by a routine running in the host machine, the following stages of processing take place:-

(a) Each word in the lexicon is selected and compared with all text words in the current window. All the words in the window that match the function word are marked with the appropriate category code from the lexicon. This concludes the dictionary lookup process and the lexicon is no longer required until the next window of text is examined.

(b) Some processing is done to associate an appropriate category code with the non-word items in the text such as punctuation marks, numbers and other symbols.

(c) Some common phrases (these will depend on the subject-area) such as "for the time being" are treated individually as being of an idiomatic nature. These are marked with a code indicating that they should be disregarded in further processing.

(d) Most of the text items not in the lexicon can be classifed by comparing the code of each known word with those of its neighbours within a region of three or four words. For example the word following a determiner must be a noun or an adjective (in either case it can be classified as part of a noun phrase) and the word following a modal must be part of a verb phrase.

Ex4.1 The powers of the court under this section relate only to regulated mortgages.

Ex4.2 The order may, if the mortgagor so requests, make provision for apportioning the money between...... .

(e) Interjections such as "if the mortgagor so requests" in Ex4.2 are dealt with by propagating the codes of preceding words over the interjection and marking the interjected words as being no longer relevant. So, in Ex.4.2, the entire sequence "may, if the mortgagor so requests, make" will be marked as one verb phrase. Bracketed expressions and expressions enclosed in quotation marks are treated similarly.

(f) Word endings such as 'ed', 'ing', 'tion', and 'ly' are tested for and given appropriate codes. These are also used in the determination of other words in their region - e.g. '--ed' words are either past tense verb parts or past participles being used adjectivally, and '--ly' words are adverbs, usually indicating the presence of another verb phrase component nearby.

For example in sentence Ex4.1 'powers', 'court' and 'section' are preceded by determiners and so are recognised as nouns; 'regulated' ends in 'ed' and since it is preceded by 'to' must be being used adjectivally, so 'mortgage' can also be classifed as a noun; 'only' is an adverb so we can classify 'relate' as the verb in the sentence.

(g) Simple coordination is dealt with here, such as verb phrase coordination, or single word coordination if recognisable.

(h) At this stage individual clauses can be identified and correspondences marked between nouns and modifying phrases. A count can then be made over each region to ensure that each clause has one and only one verb phrase, noting that relative clauses may be nested inside others and so must be done first.

(i) The more difficult coordination problems are tackled now, starting with noun phrase coordination and eventually including clause coordination, annotated lists of phrases, coordination of the type 'either .... or ....' and sentence coordination.

(j) The abstract can be completed at this stage by discarding all modifying phrases and keeping only the main verb phrase in each sentence, together with the subject.

4.2 Computational details

The lexicon is stored in the DAP in successive bit strings (DAP-Fortran logical vectors) so that it is compatible with host format. This ensures that no reformatting is required when converting between host and DAP, and that while the dictionary look-up is in progress, successive bits of a word are accessed efficiently. Some extra work is done to remove those initial letters of each word that are the same as those of the previous word. For example, if successive words in the lexicon are THEN, THERE and THEREIN, the characters stored would be THEN, RE and IN, together with an integer to indicate the number of characters to be assumed at the start of each word from the previous one (i.e. 0, 3, 5 in this instance) (see appendix 1).

The lookup is performed on text stored one character per Processing Element (P.E.), and may need to be performed on several layes of text depending on the document size. Each character in the lexicon is compared with the text characters by broadcasting to all P.E.s simultaneously. Here we can make use of the elimination of similar starting characters, as we need only do comparisons on the characters stored in the compressed lexicon.

After each layer is dealt with each word of text is represented by one 64-bit token, each bit representing a particular category of words. Some tokens may have more than one bit set; others will have no bits set if the token does not appear in the lexicon and is not a number or symbol. These tokens are then shuffled left and merged with those of successive layers so that the tokens representing 4096 words (the number of P.E.s in this DAP) may be dealt with simultaneously. This completes stage 4.(a).

The remaining processing (4.(b) to 4.(j)) reduces to logical operations on bit planes -- a method which SIMD† machines generally (and the DAP in particular) perform very efficiently compared with serial (SISD) machines. Indeed, the increase in speed obtained when comparing logical array operations on the DAP with SISD machines is much larger than the increase obtained for numerical operations. This is because SISD machines are constrained to manipulate a word at a time, whereas on the DAP the elementary operations use single bits and 'software' routines must be invoked to process numerical data, one bit-plane at a time.

Advantage can now be taken of both the bit-level nature of the machine and the interconnection network between processors which will be used for left-right communication. For example there is one bit in the encoded token which represents the lexical item "of" or "than"; another bit indicates tokens which form part of a noun phrase. In order to mark the words on either side of "of" and "than" as nouns, the OF/THAN bit-mask is shifted to the left and to the right by one place and the logical OR of these is taken. This mask is then used in a masked assignment (of the value .TRUE.) to the Noun Phrase bit mask.

---

† Single Instruction-stream, Multiple Data-stream, according to Flynn's classification (1972)

After the inital categorisation by dictionary lookup, a similar character matching process is performed (step 4.(b)) to mark with an appropriate code, the numbers and other non-letter items in the window, particularly punctuation marks. For example, one bit in the code represents sentence terminators and this is set wherever a full stop, semicolon or colon occurs.

Some of the function words in the dictionary are ambiguous in their syntactic category. For example, 'then' may be either an adverb or a conjunction, 'as' and 'that' may either be conjunctions or perform functions similar to a preposition or determiner respectively. The category of each occurrence of these words can usually be distinguished by considering their syntactic environment. For example, if a determiner follows one of these unknowns we can identify it as a conjunction. These tests, and checks for phrases of three of four words (as in sections 4.(c) and (d)) are performed in the DAP by shifting and comparing bit masks selected from the category codes. They are thus well suited to the DAP architecture. Propagation of codes and masks over variable distances is accomplished efficiently on the DAP by setting up a 'stop' mask and propagating the required mask or code, and the stop mask, in step lengths of powers of two. (see Oldfield (1982) for more details).

5. DISCUSSION

5.1 Alternatives

In trying to provide a text reduction system, I have chosen an approach suitable for SIMD processing. Having opted for such an architecture it is very difficult to make comparisons with other hardware systems, although it is clear that this is well within the capability of the DAP. This algorithm is designed for a bit-serial word-parallel architecture, since its data structures are generally logical arrays; thus it will run fast and be able to accommodate documents of a realistic size.

The performance of the syntax analysis part of this system may be compared with a functionally similar routine PARTS (Cherry (1978)) which is built into the UNIX† program STYLE (Cherry and Vestman (1979). Cherry (1982)). Cherry (1982) states that PARTS works at the rate of about 340 words per second on a PDP 11/70. From the figures given in section 6, the DAP program runs about 20 times faster than this. Disregarding I/O time (which is reasonable if I/O is done in parallel with DAP processing, as would be the case when searching a library of documents, or if the document is held in the DAP store for further processing) the process runs about 35 times faster on the DAP.

It should be noted that most of the DAP processing time is taken up in the lookup stage, which is performed five times for this sample of text (there being an average of four and a half characters per word). If this section were rewritten in APAL(ICL (1979)) a large speedup would be expected, owing to the amount of matrix indexing required in the DAP FORTRAN code.

Text parsing can also form the basis of an information retrieval system in which as well as using the associative properties of the DAP

---

† UNIX is a Trademark of Bell Laboratories.

for searching large texts for particular word patterns, more detailed queries could be defined specifying how the words are used in the target sentences. For example, to find the definition of a legal phrase a search could be made for that phrase as the subject of the verb 'mean', or the object of the verb 'refer' or 'call', etc.

Such a scheme can be compared with the STATUS system (Price, Bye and Niblett (1974)) which creates an inverted file and sets up indexes and pointers back to the original text. This consumes large amounts of batch processing time, occupies a great deal of filestore space and still requires several disc accesses to search for desired information.

5.2 Applications

Without trying to extrapolate the results to other subject areas, large volumes of legal text already exist, which are accessed frequently and which could benefit from a faster, more accurate, retrieval system. The work currently in progress is planned to cover the ground of retrieval and cross-referencing within a document.

Based on the parse information, the initial step will be to classify each sentence according to main verb type (e.g. definition, clarification, logical dependence, etc.). Given the subject and type of every sentence in a document will enable more consistent document classification to be made.

It is also hoped to be able to identify relations between the various objects in each sentence, in order to make logical deductions about the contents of a document. The abstract is a first step to this goal.

6. RESULTS

The system has been run using various extracts from the 1968 Rent Act and several other legal texts such as appeals proceedings which have a very different style. The correct subject and main verb phrase are correctly selected in about 75% of the sentences, and a reduction in text size of over 70% is normally achieved.

This accuracy compares unfavourably with the 95% claimed for PARTS by Cherry. However, when I ran STYLE on some of my legal texts, I got a success rate nearer 80%, despite the claim that the system works on 'any type of English text from business letters to technical reports'. Perhaps legal text does fall in this range. Also, in trying to relate certain clauses in complex sentences, this scheme does a deeper parse than PARTS which is only concerned with classifying each word.

The reduction is also less than the 90% reported by Edmundson (1964) in his sentence selection scheme, although it is not possible to compare the quality of the abstracts so produced. While there is still scope in this scheme for further reduction by discarding more qualifying phrases, perhaps a better abstract may be achieved by combining the two systems of sentence discards and clause/phrase discards (in remaining sentences).

The results of a run of the program with some source text taken from the 1968 Rent Act are presented in Appendix 2. The times taken in the various stages are shown below. Note that the time taken to process a document is not dependent on its content, so times are quoted per window (4096 tokens) of text.

Dap Processor Times (ms)	
Dictionary lookup	233
Code compaction	34
Syntax analysis	27
Code expansion and preparation for output	35
Total DAP MCU time	329
Host Processor Times (ms)	
Input processing (dictionary)	87
Input processing (text)	111
Output processing	16
Total host OCP time	214
Total processing	534ms

7. CONCLUSIONS

It has been shown that it is feasible to produce a reasonably accurate parse to clause level, within the given constraints. A simplistic abstract can be produced for a very large document in a few seconds, fast enough to be used interactively. Investigation is in progress on how this abstract may be used for information retrieval purposes. Another fruitful area for further work would be in extending the domain of subject-matter for which this method could be used.

8. ACKNOWLEDGEMENTS

This PhD research project is jointly funded under an industrial studentship by ICL (International Computers Limited) and the Science and Engineering Research Council of Great Britain.

Thanks are due to my supervisor, Miss E. Wilson, for help and encouragement; to several members of the Computing Laboratory for support and advice during proof-reading; and to the Computer Centre at Queen Mary College (particularly the DAP Support Unit) who have provided the essential computing services.

APPENDIX 1 - The Lexicon

The first few words in the lexicon are as follows:

Actual Word	Categories	Characters stored	Number of characters assumed	stored
A	Singular Determiner	A	0	1
Above	Preposition	BOVE	1	4
After	Prep./Conjunction	FTER	1	4
Against	Prep	GAINST	1	6
All	Noun (Adj./Pronoun)	LL	1	2
Always	Adverb	WAYS	2	4
Am	Auxiliary Verb/To Be	M	1	1
An	Sing. Det.	N	1	1
And	Coordinator	D	2	1
Another	Sing. Det.	OTHER	2	5
Any	Det.	Y	2	1
Anything	Noun	THING	3	5
Anywhere	Adverb	WHERE	3	5
Apart	Adverb	PART	1	4
Are	Verb/To Be	RE	1	2
As	Prep./Conj.	S	1	1
At	Prep.	T	1	1
Be	Infinitive/To Be	BE	0	2
Because	Conj.	CAUSE	2	5
Been	Participle/To Be	EN	2	2
Before	Conj./Prep.	FORE	2	4
Below	Prep.	LOW	2	3
Between	Alternative	TWEEN	2	5
Being	Part./To Be	ING	2	3
Bona	Idiomatic	ONA	1	3
Both	Noun (Adj.)	TH	2	2
But	Coord.	UT	1	2
By	Prep.	Y	1	1
Can	Modal	CAN	0	3
Could	Modal	OULD	1	4
Did	Aux. Verb/To Do	DID	0	3
Do	Infinitive/To Do	O	1	1
Does	Aux. Verb/To Do	ES	2	2
Each	Noun (Adj.)	EACH	0	4
Either	Alt.	ITHER	1	5
Ever	Adverb	VER	1	3
Every	Noun (Adj.)	Y	4	1
Everywhere	Adverb	WHERE	5	5
Except	Conj.	XCEPT	1	5
Fide	Idiomatic	FIDE	0	4
First	Adverb	RST	2	3
For	Prep.	OR	1	2
From	Prep.	ROM	1	3
Had	Aux. Verb	HAD	0	3
Has	Aux. Verb	S	2	1
Have	Aux. Verb	VE	2	2
He	Subject Noun	E	1	1
Her	Noun	R	2	1
Here	Abverb	E	3	1
Him	Object Noun	IM	1	2

APPENDIX 2 - Results

The following results are taken from a run where the input data was part of the 1968 Rent Act.

Appendix 2.1 Input text

This is an extract from the beginning of Part I of the Act. The Part and Section numbers and some heading text have been removed.

A tenancy under which a dwelling-house (which may be a house or part of a house) is let as a separate dwelling is a protected tenancy for the purposes of this Act unless-

(a) the dwelling-house has or had on the appropriate day a rateable value exceeding, if it is in Greater London, £400 or, if it is elsewhere, £200; or

(b) the tenancy is one with respect to which section 2 below otherwise provides; or

(c) by virtue of section 4 or section 5 below, the tenancy is for the time being precluded from being a protected tenancy by reason of the body in whom the landlord's interest is vested;

and any reference to a protected tenant shall be construed accordingly.

For the purposes of this Act, any land or premises let together with a dwelling-house shall, unless it consists of agricultural land exceeding two acres in extent, be treated as part of the dwelling-house; and for this purpose "agricultural land" has the meaning set out in paragraph (a) of section 26(3) of the General Rate Act, 1967 (which relates to the exclusion of agricultural land and premises from liability for rating).

If any question arises in any proceedings whether a dwelling-house is within the limits of rateable value in subsection (1)(a) above, it shall be deemed to be within those limits unless the contrary is shown.

A tenancy is not a protected tenancy if-

(a) under the tenancy either no rent is payable or, subject to section 7(3) below, the rent payable is less than two-thirds of the rateable value which is or was the rateable value of the dwelling-house on the appropriate day; or

(b) under the tenancy the dwelling-house is bona fide let at a rent which includes payments in respect of board, attendance or use of furniture; or

(c) subject to section 1(2) above, the dwelling-house which is subject to the tenancy is let together with land other than the site of the dwelling-house; or

(d) the dwelling-house is comprised in an agricultural holding (within the meaning of the Agricultural Holdings Act, 1948) and is occupied by the person responsible for the control (whether as tenant or as servant or agent of the tenant) of the farming of the holding.

In the following provisions of this Act, a tenancy falling within paragraph (a) of subsection (1) above is referred to as a "tenancy at a low rent".

For the purposes of paragraph (b) of subsection (1) above, a dwelling-house shall not be taken to be bona fide let at a rent which includes payments in respect of attendance or the use of furniture unless the amount of rent which is fairly attributable to attendance or use of furniture, having regard to the value of the attendance or the use to the tenant, forms a substantial part of the whole rent.

Subject to sections 4 and 5 below -

(a) after the termination of a protected tenancy of a dwelling-house the person who, immediately before that termination, was the protected tenant of the dwelling-house shall, if and so long as he occupies the dwelling-house as his residence, be the statutory tenant of it; and

(b) the provisions of Schedule 1 to this Act shall have effect for determining what person (if any) is the statutory tenant of a dwelling-house at any time after the death of a person who, immediately before his death, was either a protected tenant of the dwelling-house or the statutory tenant of it by virtue of paragraph (a) above;

and a dwelling-house is referred to as subject to a tenancy when there is a statutory tenant of it.

In paragraph (a) of subsection (1) above and in Schedule 1 to this Act, the phrase "if and so long as he occupies the dwelling-house as his residence" shall be construed as requiring the fulfilment of the same, and only the same, qualifications (whether as to residence or otherwise) as had to be fulfilled before the commencement of this Act to entitle a tenant, within the meaning of the Increase of Rent and Mortgage Interest (Restrictions) Act, 1920, to retain possession, by virtue of that Act and not by virtue of a tenancy, of a dwelling-house to which that Act applied.

A person who becomes a statutory tenant of a dwelling-house as mentioned in paragraph (a) of subsection (1) above is, in this Act, referred to as a statutory tenant by virtue of his previous protected tenancy, and a person who becomes a statutory tenant as mentioned in paragraph (b) of that subsection is, in this Act referred to as a statutory tenant by succession.

Appendix 2.2 Kernel sentences produced

This is a transcription of the program output. Any text removed is optionally replaced by '---' in the program.

A tenancy --- is a protected tenancy for the purposes --- unless (a) the dwelling-house has or had on the appropriate day a rateable value exceeding, if it is in Greater London, £400 or --- £200; or (b) the tenancy is one with respect ---; or (c) --- the tenancy is --- precluded from being a protected tenancy ---; and any reference to a protected tenant shall be construed accordingly.

--- any land or premises let together with a dwelling-house shall --- be treated as part ---; and ---

If any question arises in any proceedings whether a dwelling-house is within the limits --- above, it shall be deemed to be within those limits unless the contrary is shown.

A tenancy is not a protected tenancy if - (a) ---; or (b) -- the dwelling-house is --- let at a rent ---; or (c) -- the dwelling-house --- is let together with land other than the site ---; or (d) the dwelling-house is comprised in an agricultural holding --- and is occupied by the person responsible ---.

--- a tenancy falling --- is referred to as a "tenancy at a low rent".

--- a dwelling-house shall not be taken to be --- let at a rent --- unless the amount --- forms a substantial part of the whole rent.

--- after the termination --- the person --- shall --- be the statutory tenant of it; and (b) the provisions --- shall have effect for determining what person --- is the statutory tenant of a dwelling-house --- after the death ---; and a dwelling-house is referred to as --- when there is a statutory tenant of it.

--- the phrase "if and so long as he occupies the dwelling-house as his residence" shall be construed as requiring the fulfilment --- as had to be fulfilled before the commencement of this act ---.

A person --- is --- referred to as a statutory tenant ---, and a person --- is --- referred to as a statutory tenant ---.

9. REFERENCES

Flanders, P.M., Hunt, D.J., Reddaway, S.F. and Parkinson, D. (1977) Efficient High Speed Computing with the Distributed Array Processor, in High Speed Computer and Algorithm Organisation, (D.J. Kuck, D.H. Lawrie and A.H. Sameh, eds.), Academic Press, pp. 113-128.

Parkinson, D. (1980) The Distributed Array Processor, *IUCC Bulletin*, **2**, pp. 119-121.

Wilson, E. and Ball, A.G. (1974) The syntactic analysis of formal English text to create a structured data-base, Report to British Library - Project S1/82/05, Computing Lab. UKC.

Luhn, H.P. (1958) Automatic creation of literature abstracts, *IBM Journal Res. Dev.*, **2**, pp. 159-165.

Edmundson, H.P. (1964) Problems in Automatic abstracting, *Comm. ACM*, **7**, pp. 259-263.

Edmundson, H.P. (1969) New methods in automatic abstracting, *J. ACM*, **16**, pp. 264-285.

Mathis, B., Rush, J.E. and Young, C.E. (1973) Improvement of automatic abstracts by the use of structural analysis, *J. of the American Soc. for Info. Sci.*, **24**, pp. 101-109.

Flynn, M. (1972) Some computer organisations and their effectiveness, *IEEE Trans. Comp.*, *C*-**21**, pp. 948-960

Oldfield, D.E. (1982) The propagation problem, DAPSU newsletter, Queen Mary College, London, Dec.

Cherry, L.L. (1978) PARTS - A system for assigning word classes to English text, Comp. Sci. Tech. Report 81, Bell Lab., Murray Hill, NJ.

Cherry, L.L. and Vestman, W. (1979) Writing tools - The STYLE and DIRECTION programs, Comp. Sci. Tech. Report 91, Bell Lab., Murray Hill, N.J.

Cherry, L.L. (1982) Writing tools, *IEEE Trans. Comm.*, *Com*-**30**, pp. 100-105.

ICL (1979) DAP: APAL Language, TP6919, International Computers Ltd., London.

Price, N.H., Bye, C. and Niblett, B. (1974) On-line searching of Council of Europe conventions and agreements: a study in bilingual document retrieval, Harwell Report AERE-R7673, HMSO, London.

SPARSE MATRIX VECTOR MULTIPLICATION ON THE DAP

R.H. Barlow, D.J. Evans and J. Shanehchi

(Department of Computer Studies, Loughborough University of Technology, Loughborough)

1. INTRODUCTION

Many important algorithms require repeated multiplication of the type AX=b where A is a sparse matrix and X is a dense vector. Several authors have considered schemes for the efficient implementation of sparse matrix operations on the ICL DAP for cases when the size of the matrix exceeds the size of the DAP array (64 by 64). They have however considered the case of unstructured sparse matrices; that is the null entries in the matrix can be anywhere. Morjaria and Mackinson (1981) divided the matrix into 64 by 64 blocks and stored corresponding non-null elements of blocks at the same processor. Parkinson (1981) treated the DAP array as a linear line of 4096 processors and then stored one compacted row of A per processor. The unstructured sparsity of the system means that both schemes have a variable number of elements stored at a processor and that each component of the vector x has to be made available to each processor. Both these features give rise to significant losses of potential power.

Here we consider structured sparse matrices: in particular banded systems where bands parallel to the diagonal are considered full or empty. Such banded systems arise naturally in structural analysis problems. In fact the original data is often manipulated (permutated) to yield a more banded form. After outlining the method and results obtained on the DAP we show how one of the parts of this method can be incorporated into Parkinson's scheme leading to a significant increase in speed for this latter unstructured sparse matrix method.

2. THE METHOD

The ICL-DAP is a parallel machine utilising Single Instruction Multiple Data type parallelism (Flanders, Hunt, Reddaway and Parkinson (1977)). It consists of a control processor and a 64X64 matrix of slave processors. The slaves are interconnected so that they can in parallel communicate with their four nearest neighbours. The array of slaves can logically be considered as a line of 4096 processors each of which can communicate with its left or right neighbour: communication in this representation takes almost twice as long due to lack of a direct hardware connection between the edge processors. For clarity in the following discussion an east or west shift will be called a column shift. The ICL-DAP has some vector broadcast facilities: copying for example a single column to all columns in two operations. This is useful for replicating a vector.

The size of the matrix and vector are denoted by n and we assume $64 \leq n \leq 4096$. To make use of the full processing power and local connectivity of the ICL DAP two techniques are used. Firstly several bands of A are stored on each DAP storage plane so that the number of processors that can in parallel compute partial results is maximised. Secondly storage for each band starts at a new column of the DAP: the intention being to use column shift operations to add up corresponding partial results rapidly. A band of the matrix is stored one element per processor and each band occupies n processors irrespective of whether the band has this number of elements. Fig. 1 illustrates one implementation of this storage scheme. When no more complete bands can be stored on a plane then storage on the next DAP plane starts. This second technique limits slightly the packing of bands on planes since some processors have no components of A or X allocated to them.

PLANE 1 OF A

a_{11}	a_{99}	–	a_{89}	–	a_{79}	–	a_{69}
a_{22}	$a_{10,10}$	a_{12}	$a_{9,10}$	–	$a_{8,10}$	–	$a_{7,10}$
a_{33}	$a_{11,11}$	a_{23}	$a_{10,11}$	a_{13}	$a_{9,11}$	–	$a_{8,11}$
a_{44}	$a_{12,12}$	a_{34}	–	a_{24}	–	a_{14}	–
a_{55}	–	a_{45}	–	a_{35}	–	a_{25}	–
a_{66}	–	a_{56}	–	a_{46}	–	a_{36}	–
a_{77}	–	a_{67}	–	a_{57}	–	a_{47}	–
a_{88}	–	a_{78}	–	a_{68}	–	a_{58}	–
Main Diagonal		UB1		UB2		UB3	

PLANE 2 OF A

–	$a_{5,9}$	–	$a_{4,9}$	a_{21}	$a_{10,9}$	a_{31}	$a_{11,9}$
–	$a_{6,10}$	–	$a_{5,10}$	a_{32}	$a_{11,10}$	a_{42}	$a_{12,10}$
–	$a_{7,11}$	–	$a_{6,11}$	a_{43}	$a_{12,11}$	a_{53}	–
–	$a_{8,12}$	–	$a_{7,12}$	a_{54}	–	a_{64}	–
a_{15}	–	–	–	a_{65}	–	a_{75}	–
a_{26}	–	a_{16}	–	a_{76}	–	a_{86}	–
a_{37}	–	a_{27}	–	a_{87}	–	a_{97}	–
a_{48}	–	a_{38}	–	a_{98}	–	$a_{10,8}$	–
UB4		UB5		LB1		LB2	

Upper Diagonal Bands — Lower Diagonal Bands

PLANE 3 OF A

a_{41}	$a_{12,9}$	a_{51}	–	a_{61}	–	–	–
a_{52}	–	a_{62}	–	a_{72}	–	–	–
a_{63}	–	a_{73}	–	a_{83}	–	–	–
a_{74}	–	a_{84}	–	a_{94}	–	–	–
a_{85}	–	a_{95}	–	$a_{10,5}$	–	–	–
a_{96}	–	$a_{10,6}$	–	$a_{11,6}$	–	–	–
$a_{10,7}$	–	$a_{11,7}$	–	$a_{12,7}$	–	–	–
$a_{11,8}$	–	$a_{12,8}$	–	–	–	–	–
LB3		LB4		LB5			

Lower Diagonal Bands (cont.)

PLANE OF X (INITIALLY)

x_1	x_9	–	–	–	–	–	–
x_2	x_{10}	–	–	–	–	–	–
x_3	x_{11}	–	–	–	–	–	–
x_4	x_{12}	–	–	–	–	–	–
x_5	–	–	–	–	–	–	–
x_6	–	–	–	–	–	–	–
x_7	–	–	–	–	–	–	–
x_8	–	–	–	–	–	–	–
Vector X							

Fig. 1 Storage of A and X (1st implementation)

If nbands, ncols, npack and nplane denote respectively the number of non-empty bands of the matrix A, the number of columns required to store each band, the number of bands that can be stored on each storage plane and finally the number of storage planes required to store all the bands then

$$\text{ncols} = \text{INT}^{*}(\text{n}-1,64)+1; \qquad \text{npack} = \text{INT}(64,\text{ncols});$$

$$\text{nplane} = \text{INT}(\text{nbands}-1,\text{npack})+1$$

Two implementations were considered and these are now described in the order in which they were developed.

To make the exposition clearer several figures illustrate the storage of A and X for the case of an 8 by 8 ICL DAP and a matrix of size n=12 with 11 non-empty bands: the diagonal, the first five super-diagonals and the first five lower diagonals (Fig. 1).

2.1 Implementation 1

The intention was to minimise the effort in making components of X available at processors storing components of A requiring their use. Since component X(i) is only required by elements of A on column i this can be achieved by storing all components on column i of A and the component i of X at row i of the DAP array. Taking into account that a band of A and the whole of X will require more than one column for storage the resulting initial storage of X and A is as shown in Fig. 1. The vector X can be replicated over the plane using either a number, equal to the number of columns occupied by X, of column broadcasts or by a number of column shift operations. Either option is highly parallel and for a given size matrix the implementation chose the optimum. The result of replicating X is shown in Fig. 2 and comparison of this with Fig. 1 illustrates that each component of X is present at processors storing corresponding components of A.

x_1	x_9	x_1	x_9	x_1	x_9	x_1	x_9
x_2	x_{10}	x_2	x_{10}	x_2	x_{10}	x_2	x_{10}
x_3	x_{11}	x_3	x_{11}	x_3	x_{11}	x_3	x_{11}
x_4	x_{12}	x_4	x_{12}	x_4	x_{12}	x_4	x_{12}
x_5	–	x_5	–	x_5	–	x_5	–
x_6	–	x_6	–	x_6	–	x_6	–
x_7	–	x_7	–	x_7	–	x_7	–
x_8	–	x_8	–	x_8	–	x_8	–

Fig. 2 Replicated vector X

---

* INT(a,b) denotes the integer part of a divided by b.

Component multiplication, once for each storage plane, now yields all partial products. However these partial products are not in the positions where they can be easily totalled together. Thus we require that the partial results on different planes are aligned so that planes of partial results can be added together, to yield a single plane of npack sets partial results, and so that the bands of partial results on this resulting plane can be totalled using ln2(npack) sets of (parallel) column shifts and addition. This alignment is in fact not difficult to achieve since the non-matching of the original partial results is highly regular. Partial results arising from a band m below (above) the diagonal band must be shifted m places right (left). Once this has been done all results at a given processor contribute to the same result component and partial results corresponding to the same final component lie on the same row.

Since several bands can be stored on each plane the results on a plane must be shifted several times. However since bands are stored in increasing offset from the diagonal (or equivalently increasing number of shifts required) then on each plane each band, except the first, needs shifting by only the difference between the band offsets. Obviously the shifting process must be repeated twice over on the plane that contains a mixture of upper and lower bands.

An estimate of the performance of the method can be found by assuming that all bands lie within distance m of the diagonal and that in order to bring the partial products on each plane into the corresponding positions requires on average m/2 left and/or m/2 right shifts on each plane. Using the following DAP timings for 32 bit representation of real numbers:

parallel add real numbers ~ 175 microseconds
parallel multiply real numbers ~ 275 microseconds
parallel shift real numbers ~ 25 microseconds per displacement
parallel broadcast a column ~ 30 microseconds

then the time for matrix vector multiplication is

T(n,m) < ncol*30+nplane*275+m/2*nplane*25+(nplane-1)*175+ln2(nband)*175+(64-ncol)*25

Taking n=1024 and m=11 the time taken is

T(1024,11) = 0.5+1.6+0.9+0.9+0.7+1.2 = 5.8 milliseconds

Results for vector matrix multiplication of a number of banded matrices are shown in Table 1. It is worthwhile remembering that vector matrix multiplication of a full 64 by 64 matrix takes 0.65 milliseconds on the DAP. For large banded systems the advantage of this representation can be seen.

2.2 Implementation 2

The previous implementation requires that each band be shifted. While within a plane the effect of shifting the previous band reduces the shifting required of subsequent bands this still means that the initial band on each plane may require significant shifting.

Table 1

Results on the 1st implementation
Speedup compared with ICL-2980

Size \ No. of Bands	3		7		11		15		23		31	
	Time	Speed up	T	SU	T	SU	T	SU	T	SU	T	SU
64	1.7	0.48	2.37	0.77	3.1	0.84	3.72	0.93	5.1	1.02	6.42	1.14
128	1.96	0.79	2.97	1.19	4.1	1.24	4.71	1.42	6.54	1.64	7.89	1.93
256	2.21	1.27	3.26	2.01	4.4	2.25	5.1	2.67	7.1	3	8.39	3.63
512	2.21	2.48	3.66	3.59	4.94	4.05	5.6	4.77	7.61	5.63	9.54	6.4
1024	2.53	4.29	3.79	6.82	5.1	7.82	6.34	8.41	8.94	9.64	11.61	10.58
2048	2.53	8.48	4.41	11.6	6.28	12.68	8.18	13.34	12	14.69	16.1	15.36
4096	2.3	18.69	5.3	19.49	8.38	19.87	11.52	19.39	17.97	20.2	24.64	20.44

The shifting can be reduced when an alternate representation is used whereby the partial results are generated in their final corresponding positions but the initial X for each band must be shifted. Fig. 3 illustrates the storage planes of A for this implementation. The storage of X for the first two storage planes of A is shown in Fig. 4. Having shifted the copy of X for the bands stored on one plane, then since the bands are stored in increasing offset from the diagonal (or equivalently increasing amount of shift required) all the bands on the next plane require more shifting of X than those on the current plane. Thus the copy of X shifted for the current plane can be used as the starting copy for shifting on the next plane. Obviously, this cannot be used for the first plane or for the first plane of bands requiring shifts in the opposite direction (one must restart from the original X).

The only complication arising from using the previous X as the starting point for the new plane is that relative shift X for bands on two successive planes does not now increase with increasing band offset. To minimise the shifting within a plane the various shifts relative to the previous plane are sorted so that except for the first shift on a plane, by an amount equal to the minimum shift, all subsequent shifts within a plane are with respect to the difference between the required and current subtotal of shifts carried out.

In circumstances when the full bands are regularly spaced the method is very efficient. Consider then the example of n=1024 and m=11. The first plane has bands with offsets 0,1,2 and 3 while the second plane has bands with offsets 4,5,6 and 7. Thus the X on the whole of the second plane can be obtained by shifting the plane of X used for the corresponding first plane of A once and once only by 4 positions.

PLANE 1 OF A

a_{11}	a_{99}	a_{12}	$a_{9,10}$	a_{13}	$a_{9,11}$	a_{14}	$a_{9,12}$
a_{22}	$a_{10,10}$	a_{23}	$a_{10,11}$	a_{24}	$a_{10,12}$	a_{25}	–
a_{33}	$a_{11,11}$	a_{34}	$a_{11,12}$	a_{35}	–	a_{36}	–
a_{44}	$a_{12,12}$	a_{45}	–	a_{46}	–	a_{47}	–
a_{55}	–	a_{56}	–	a_{57}	–	a_{58}	–
a_{66}	–	a_{67}	–	a_{68}	–	a_{69}	–
a_{77}	–	a_{78}	–	a_{79}	–	$a_{7,10}$	–
a_{88}	–	a_{89}	–	$a_{8,10}$	–	$a_{8,11}$	–
Main Diagonal		UB1		UB2		UB3	

PLANE 2 OF A

a_{15}	–	a_{16}	–	–	a_{98}	–	a_{97}
a_{26}	–	a_{27}	–	a_{21}	$a_{10,9}$	–	$a_{10,8}$
a_{37}	–	a_{38}	–	a_{32}	$a_{11,10}$	a_{31}	$a_{11,9}$
a_{48}	–	a_{49}	–	a_{43}	$a_{12,11}$	a_{42}	$a_{12,10}$
a_{59}	–	$a_{5,10}$	–	a_{54}	–	a_{53}	–
$a_{6,10}$	–	$a_{6,11}$	–	a_{65}	–	a_{64}	–
$a_{7,11}$	–	$a_{7,12}$	–	a_{76}	–	a_{75}	–
$a_{8,12}$	–	–	–	a_{87}	–	a_{86}	–
UB4		UB5		LB1		LB2	

Upper Diagonal Bands (UB1–UB5); Lower Diagonal Bands (LB1–LB2)

PLANE 3 OF A

–	a_{46}	–	a_{95}	–	a_{94}	–	–
–	$a_{10,7}$	–	$a_{10,6}$	–	$a_{10,5}$	–	–
–	$a_{11,8}$	–	$a_{11,7}$	–	$a_{11,6}$	–	–
a_{41}	$a_{12,9}$	–	$a_{12,8}$	–	$a_{12,7}$	–	–
a_{52}	–	a_{51}	–	–	–	–	–
a_{63}	–	a_{62}	–	a_{61}	–	–	–
a_{74}	–	a_{73}	–	a_{72}	–	–	–
a_{85}	–	a_{84}	–	a_{83}	–	–	–
LB3		LB4		LB5			

Fig. 3 Storage of A (2nd implementation)

x_1	x_9	x_2	x_{10}	x_3	x_{11}	x_4	x_{12}
x_2	x_{10}	x_3	x_{11}	x_4	x_{12}	x_5	–
x_3	x_{11}	x_4	x_{12}	x_5	–	x_6	–
x_4	x_{12}	x_5		x_6	–	x_7	–
x_5	–	x_6	–	x_7	–	x_8	–
x_6	–	x_7	–	x_8	–	x_9	–
x_7	–	x_8	–	x_9	–	x_{10}	–
x_8	–	x_9	–	x_{10}	–	x_{11}	–

x_5	–	x_6	–	x_1	x_1	x_1	x_9
x_6	–	x_7	–	x_2	x_{10}	x_2	x_{10}
x_7	–	x_8	–	x_3	x_{11}	x_3	–
x_8	–	x_9	–	x_4	–	x_4	–
x_9	–	x_{10}	–	x_5	–	x_5	–
x_{10}	–	x_{11}	–	x_6	–	x_6	–
x_{11}	–	x_{12}	–	x_7	–	x_7	–
x_{12}	–	–	–	x_8	–	x_8	–

Fig. 4 Storage of X for the first two planes of A (2nd implementation)

Results shown in Table 2 indicate that this method is more efficient than our first implementation. The improvement due to packing bands onto planes can be seen by comparing results for a given size with the corresponding results for the system of size n=4096, since for this latter case no packing occurs.

Table 2

Results on the 2nd implementation
Speedup compared with ICL-2980

Size \ No. of Bands	3		7		11		15		23		31	
	Time	Speed up	T	SU	T	SU	T	SU	T	SU	T	SU
64	1.53	0.54	2.08	0.88	2.63	0.99	3.18	1.08	4.28	1.22	5.38	1.36
128	1.68	0.93	2.57	1.38	3.50	1.45	4.06	1.65	5.65	1.9	6.74	2.25
256	1.72	1.63	2.66	2.46	3.70	2.67	4.26	3.19	6.0	3.55	6.98	4.36
512	1.8	3.04	2.84	4.64	4.03	4.96	4.48	5.97	5.5	7.79	6.19	9.87
1024	1.95	5.57	3.04	8.5	4.04	9.87	4.71	11.31	6.24	13.81	7.76	15.82
2048	2.1	10.22	3.74	13.68	5.24	15.19	6.74	16.19	9.7	18.18	12.74	19.41
4096	2.32	18.53	5.31	19.46	8.28	20.1	11.27	19.83	17.23	21.1	23.14	21.77

3. IMPROVEMENTS TO THE METHOD OF PARKINSON

It is now worthwhile examining the relationship of this method to that of Parkinson. As previously stated his method for arbitrary sparse systems involves each processor storing all the non-null elements of a row. It follows that once the partial results are available all those corresponding to the same result component lie at the same processor. Thus parallel summation over the planes yields the results vector.

For arbitrary sparse systems then each plane storing A requires an arbitrary collection of components of X. This routine is costly and Parkinson quotes times of 11 milliseconds or 20*n microseconds per plane for respectively a low level permutation algorithm and sequential broadcasting of each component of X. Parkinson was aware that these costs could be reduced for example by calculating the minimum and maximum column number of any plane or taking account of any regular structure in the system. However the points to be made below apply to all the implementations in his solution.

Comparing this scheme with ours it is simple to envisage a useful improvement that can be made for systems of size <2048. That is that each plane of storage should be packed as in our example. This reduces the number of required planes and by a corresponding amount the time for both routing the X(i) to all required positions, and then computing the partial results on a plane and summing them into the results of the previous planes. However there is a small penalty to be paid since now partial results exist on the one final plane and these must be totalled.

Thus in general one can expect the computation to be reduced by a factor npack: this improvement should be similar to that occurring for the banded representation discussed above.

It is interesting to compare the storage representation of Parkinson with those of our second implementation. They are very similar. Thus if the implementation above did not pack the plane then each processor stores a row of matrix A. However the order of storage of the elements in a row is slightly different from that of Parkinson. Thus the implementation above starts the row storage at the diagonal element and when no more upper diagonal elements remain to be stored starts the storage at the lower diagonal elements. If the processors of the DAP array could index separately into the planes the two representations could be collapsed into one. This equivalence holds as well when planes are packed. However the equivalence holds only if the bands are full since the banded representation above allocates storage space even if an element is null.

It is interesting to compare the cost of using an unstructured sparse representation when the matrices are banded. Thus assuming a matrix of size n=1024 with semi-bandwidth 11 (that is 23 bands) then storage requires 6 planes and thus 6 planar multiplication operations and 5 planar add operations are required for both schemes. The real difference occurs when moving X to required locations. Thus Parkinson's scheme required 6*11=66 milliseconds using the permutation routing while the banded representation requires only 4*25=100 microseconds for the first plane, 1*25 microseconds for the next two planes containing super-diagonals, 100 microseconds for the fourth plane that holds the first set of lower diagonal bands followed by 1*25 microseconds routing times for the final two planes holding more lower-diagonal bands. Thus the routing costs are reduced by a factor of 200 and the banded representation executes 24 times faster than the unstructured representation.

4. CONCLUSION

The results on matrix vector multiplication of banded systems are encouraging. It should be remembered that the processors of the ICL DAP array have no floating point hardware and for purely floating point operation then the array is normally only about 5 to 15 times as fast as the ICL 2980. However while integer, boolean or fixed point operations would yield significantly better results the cost of routing the data within the array would become the dominant cost of the operation. It follows that a scheme that minimises routing becomes then crucial.

The schemes for banded sparse systems are successful because they incorporate two techniques: firstly minimising routing costs and secondly by utilising as many processors in the DAP array as is consistent with minimal routing. The packing technique that allows most processors to be utilised can be used to significant effect in the method of Parkinson for arbitrary sparse systems.

In the text it was demonstrated that techniques, such as the one presented, that took account of the structured sparsity of systems could be twenty or so times more efficient than solutions that ignored this. This result emphasises the fact that for array processors where processing is cheap relative to routing that it may be worthwhile to take an arbitrary sparse system and try to produce from this an equivalent banded, or largely banded system.

To examine this possibility we are currently using the banded representation outlined above in some iterative schemes where the matrix vector operation is repeated numerous times. Coupled with this we are investigating the cost of an algorithm to regularise the bandwidth of arbitrary sparse systems.

REFERENCES

Morjaria, M. and Makinson, G.J. (1981) Operations with sparse matrices on the ICL Distributed Array Processor, Mathematical Institute, University of Kent at Canterbury, Kent, England.

Parkinson, D. (1981) Sparse matrix vector multiplication on the DAP, DAP Support Unit, Queen Mary College, London, England.

Flanders, P.M., Hunt, D.J., Reddaway, S.J. and Parkinson, D. (1977) Efficient High Speed Computing with the Distributed Array Processor, in High Speed Computer and the Algorithm Organisation, (D.J. Kuck, D.H. Lawrie and A.H. Sameh, eds.), Academic Press, pp. 113-128.

UNSTRUCTURED SPARSE MATRIX VECTOR MULTIPLICATION ON THE DAP

M. Morjaria and G.J. Makinson

(Mathematical Institute, University of Kent at Canterbury)

1. INTRODUCTION

The computation of the matrix product $A\underline{x} = \underline{y}$ where A is a sparse matrix is a fundamental step in many algorithms. In some cases, for example in linear programming, the dimensions of A may be as high as 10,000, A may be unstructured and the density of the non zero elements is often as low as 0.1%.

This paper examines two algorithms which can be used to carry out the matrix vector product when A is sparse and unstructured. Algorithm "2D Array" in Morjaria and Makinson (1981), is based on partitioning the matrix into 64 x 64 blocks. It packs the non null coefficients of each block into matrix planes retaining the positions of the non null coefficients within the blocks by mapping onto the 2 dimensional array of processing elements. Algorithm "Long Vector", in Parkinson (1981), treats the DAP array as linear array of 4096 processors and stores one row of A in compact form per processor.

Both algorithms attempt to take advantage of the parallel processing power of the DAP to perform the arithmetic but the DAP's ability to handle data using global broadcasting techniques and logical masking operations performed in parallel turns out to be just as important.

2. ALGORITHM "2D ARRAY"

Storage

The storage technique devised for this algorithm exploits the two dimensional plane geometry of the processors such that the indices of the non zero values are associated with the position of the processing elements.

Consider the matrix A, which may be rectangular, to be partitioned into 64 x 64 blocks. The matrix is bordered with zeros if necessary and this incurs minimal penalty. The blocks are numbered as in normal matrix notation and placed on top of one another in the following order:

Block (1,1), Block (1,2), ..., Block (1,N), Block (2,1), ..., Block (2,N), ..., Block (M,1), ..., Block (M,N).

In DAP Fortran the set of n matrices would be declared by Block (,,n) and to index these matrices the kth matrix of the set would be called by Block (,,k).

The blocks are compressed together in order to squeeze out as many zeros as possible and a list is set up within each processor (in the same order as the non zeros in that processor) of the row - column block indices of the blocks from which the non zeros have come.

If the non zeros appear in random positions then this compression process will be highly successful in packing the information into as few planes as possible. For particular structured matrices the partitioning could result in a wasteful use of the store when the same position is occupied in very many blocks.

For example consider a 5 x 5 sparse matrix. This is extended to a 6 x 6 matrix.

	Col Block 1		Col Block 2		Col Block 3		
	5.0	0.0	0.0	1.0	0.0	0.0	Row Block 1
	2.0	8.0	0.0	0.0	0.0	0.0	
	1.0	0.0	3.0	1.0	0.0	0.0	Row Block 2
	0.0	7.0	0.0	1.0	0.0	0.0	
	0.0	0.0	0.0	0.0	4.0	0.0	Row Block 3
Border	0.0	0.0	0.0	0.0	0.0	0.0	

Fig. 1

When mapped on a 2 by 2 DAP, this is represented in the processor stores by four matrix planes as follows:

$$A(,,1) = \begin{bmatrix} 5.0 & 1.0 \\ 2.0 & 8.0 \end{bmatrix} \quad Arow(,,1) = \begin{bmatrix} 1 & 1 \\ 1 & 1 \end{bmatrix} \quad Acol(,,1) = \begin{bmatrix} 1 & 2 \\ 1 & 1 \end{bmatrix}$$

$$A(,,2) = \begin{bmatrix} 1.0 & 1.0 \\ 0.0 & 7.0 \end{bmatrix} \quad Arow(,,2) = \begin{bmatrix} 2 & 2 \\ 0 & 2 \end{bmatrix} \quad Acol(,,2) = \begin{bmatrix} 1 & 2 \\ 0 & 1 \end{bmatrix}$$

$$A(,,3) = \begin{bmatrix} 3.0 & 0.0 \\ 0.0 & 1.0 \end{bmatrix} \quad Arow(,,3) = \begin{bmatrix} 2 & 0 \\ 0 & 2 \end{bmatrix} \quad Acol(,,3) = \begin{bmatrix} 2 & 0 \\ 0 & 2 \end{bmatrix}$$

$$A(,,4) = \begin{bmatrix} 4.0 & 0.0 \\ 0.0 & 0.0 \end{bmatrix} \quad Arow(,,4) = \begin{bmatrix} 3 & 0 \\ 0 & 0 \end{bmatrix} \quad Acol(,,4) = \begin{bmatrix} 3 & 0 \\ 0 & 0 \end{bmatrix}$$

Fig. 2

The vector $\underline{x}$ to be multiplied may be represented in a similar way

$$\text{Let } \underline{x} = \begin{bmatrix} 9.0 \\ 1.0 \\ 0 \\ 6.0 \\ 11.0 \end{bmatrix}$$

This is first extended to 6 x 1 and then compressed and stored as the two vectors

$$[9.0 \quad 1.0] \quad \text{and} \quad [11.0 \quad 6.0]$$

with row block indices

$$[1 \quad 1] \quad \text{and} \quad [3 \quad 2]$$

The most important characteristic of this storage pattern is that the position of non zero element in A or $\underline{x}$ is determined implicitly from the co-ordinates of the processor where it is stored and the address of its block (i.e. row-column block indices), which is stored with it.

3. BASIC STEPS OF ALGORITHM "2D ARRAY" FOR MATRIX VECTOR MULTIPLICATION

(1) Broadcast the row block indices of $\underline{x}$ along the column highways to all the processors.

(2) Compare the column block indices of the elements of A in each plane with the transmitted row block indices of $\underline{x}$. Set up a logical mask on each plane to indicate where the indices are the same. This mask indicates which non zeros of $\underline{x}$ are required and also the non zeros from A with which they will need to be multiplied.

(3) Multiply pairs from $\underline{x}$ and A. The number of multiplication steps needed is dependent on the maximum number of compressed elements of $\underline{x}$ or A under a processing element. The products are stored under the same processing element. It is possible to shift non zeros to other vacant positions to reduce the number of planes needed to store the coefficients and this would reduce the number of plane operations required.

There would however be an overhead involved in the transfer and in the relocation of the product elements.

(4) The products formed in (3) have now to be summed in groups. However the numbers to be summed are all within the same row of processors. The addition stage is performed in two steps. It is carried out by collecting the entries according to their row block indices. A search is made through the stored planes and the product elements with the same row block indices are added by parallel plane additions. Then by means of a set of row sums across the processors the final entries in the matrix vector

product are obtained. Their respective row block indices are also obtained by means of corresponding Boolean plane operations. The matrix vector product is obtained in full and not packed irrespective of the way the vector in the product was stored, packed or unpacked.

4. ILLUSTRATED EXAMPLE $A\underline{x} = \underline{y}$

With A and $\underline{x}$ as shown in Section 2 the multiplication proceeds as follows.

Each block vector of $\underline{x}$ is expanded by rows to form a matrix plane and the block indices are treated likewise.

Thus MATX(,,1) $= \begin{bmatrix} 9.0 & 1.0 \\ 9.0 & 1.0 \end{bmatrix}$ with row block indices $\begin{bmatrix} 1 & 1 \\ 1 & 1 \end{bmatrix}$

and MATX(,,2) $\begin{bmatrix} 11.0 & 6.0 \\ 11.0 & 6.0 \end{bmatrix}$ with row block indices $\begin{bmatrix} 3 & 2 \\ 3 & 2 \end{bmatrix}$

The element by element multiplication of each A(,,i) i = 1, ..4 (denoted by * in DAP Fortran) with each of the matrices MATX(,,1) and MATX(,,2) is carried out if a matching between a column block index of A and a row block index of MATX occurs for a particular pair of matrices. Only the coefficients in the positions corresponding to the matchings are multiplied.

Following successful matching it is the row block indices of A (not the column block indices) in the matching positions which are retained.

The result for this example is the following set of product planes, together with their respective row block index planes.

PRODUCT PLANES ROW BLOCK INDEX PLANES

A(,,1)* MATX(,,1) $= \begin{bmatrix} 45.0 & - \\ 18.0 & 8.0 \end{bmatrix}$ $\begin{bmatrix} 1 & - \\ 1 & 1 \end{bmatrix}$

A(,,2)* MATX(,,1) $= \begin{bmatrix} 9.0 & - \\ - & 7.0 \end{bmatrix}$ $\begin{bmatrix} 2 & - \\ - & 2 \end{bmatrix}$

PRODUCT PLANES ROW BLOCK INDEX PLANES

$$A(,,1)*MATX(,,2) = \begin{bmatrix} - & 6.0 \\ - & - \end{bmatrix} \qquad \begin{bmatrix} - & 1 \\ - & - \end{bmatrix}$$

$$A(,,2)*MATX(,,2) = \begin{bmatrix} - & 6.0 \\ - & - \end{bmatrix} \qquad \begin{bmatrix} - & 2 \\ - & - \end{bmatrix}$$

$$A(,,3)*MATX(,,2) = \begin{bmatrix} - & - \\ - & 6.0 \end{bmatrix} \qquad \begin{bmatrix} - & - \\ - & 2 \end{bmatrix}$$

$$A(,,4)*MATX(,,2) = \begin{bmatrix} 44.0 & - \\ - & - \end{bmatrix} \qquad \begin{bmatrix} 3 & - \\ - & - \end{bmatrix}$$

By carrying out parallel element by element additions across these planes it is possible to produce planes representing the sum of all elements with row block index I, I = 1,2,3. Again addition need only be performed if a test on the indices indicates a non null plane.

This gives I = 1 $\begin{bmatrix} 45.0 & 6.0 \\ 18.0 & 8.0 \end{bmatrix}$

I = 2 $\begin{bmatrix} 9.0 & 6.0 \\ 0 & 13.0 \end{bmatrix}$

I = 3 $\begin{bmatrix} 44.0 & 0 \\ 0 & 0 \end{bmatrix}$

A row sum operation can now be performed along the rows of each of the planes and this gives

I = 1 $\begin{bmatrix} 51.0 \\ 26.0 \end{bmatrix}$ I = 2 $\begin{bmatrix} 15.0 \\ 13.0 \end{bmatrix}$ I = 3 $\begin{bmatrix} 44.0 \\ 0 \end{bmatrix}$

corresponding to the partitions of the vector result of the matrix vector multiplication and I is the corresponding row block index.

the result of the original 5 x 5 matrix vector multiplication is $\underline{y}$

$$\underline{y} = \begin{bmatrix} 51.0 \\ 26.0 \\ 15.0 \\ 13.0 \\ 44.0 \end{bmatrix}$$

5. ALGORITHM "LONG VECTOR" IN BASE FORM

Storage

We assume that the vector and matrix are of order N where N≤4096. We further assume that the maximum number of elements in any row is <50 so that the matrix A can be stored one row per processor with each row being defined by number pairs (non zero, J) where the non zero coefficient is in column J of A. Therefore, the number of DAP planes required will be dependent on the maximum number of non zeros in any row. It is further assumed that processor i contains x_i and will receive y_i.

To illustrate the above the matrix A of Fig. 1 with its corresponding column indices is shown in Fig. 3 again assuming a 2 x 2 DAP.

$$A(,,1) = \begin{bmatrix} 5.0 & 1.0 \\ 2.0 & 7.0 \end{bmatrix} \qquad Acol(,,1) = \begin{bmatrix} 1 & 1 \\ 1 & 2 \end{bmatrix}$$

$$A(,,2) = \begin{bmatrix} 1.0 & 3.0 \\ 8.0 & 1.0 \end{bmatrix} \qquad Acol(,,2) = \begin{bmatrix} 4 & 3 \\ 2 & 4 \end{bmatrix}$$

Fig. 3

6. BASIC STEPS OF ALGORITHM "LONG VECTOR"

(1) Use the micro code permutation routine PERMUTE(X,INDEX), which returns a matrix variable such that the ith element is X(INDEX(i)), to obtain a matrix plane W with permuted non zero elements of $\underline{x}$.

(2) Perform an element by element multiplication between A(,,I) and corresponding W. The products are stored under the same processing element.

(3) Finally, add all the resultant matrix planes formed in (2) to obtain the result vector $\underline{y}$.

Various improvements can be made to this base version of the algorithm to take account of the variable amount of sparsity per row and also the overall degree of sparsity in the matrix.

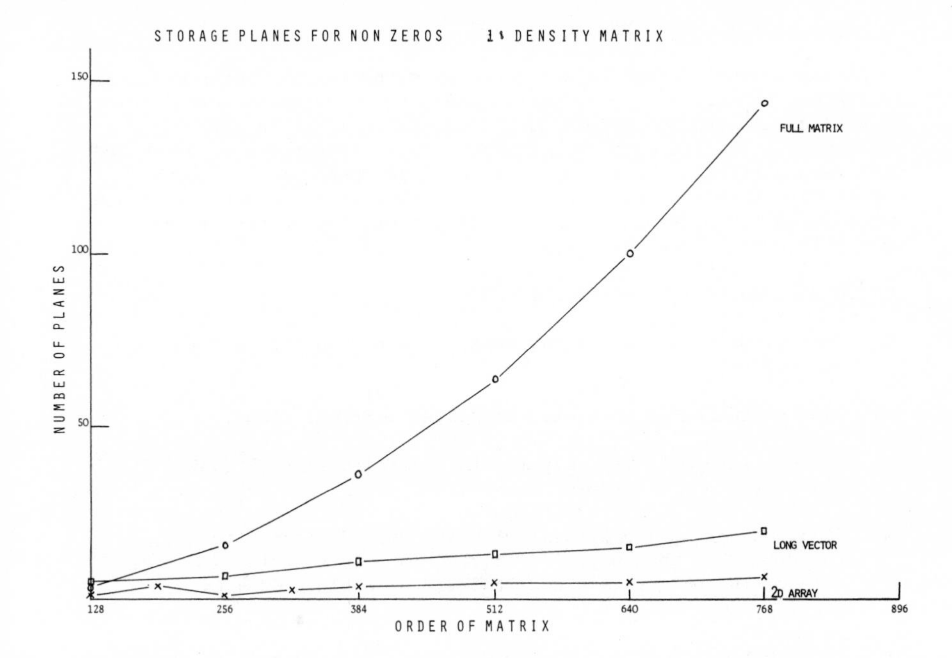
STORAGE PLANES FOR NON ZEROS 1% DENSITY MATRIX
NUMBER OF PLANES
150
100
50
FULL MATRIX
LONG VECTOR
2D ARRAY
128
256
384
512
640
768
896
ORDER OF MATRIX

Fig. 4

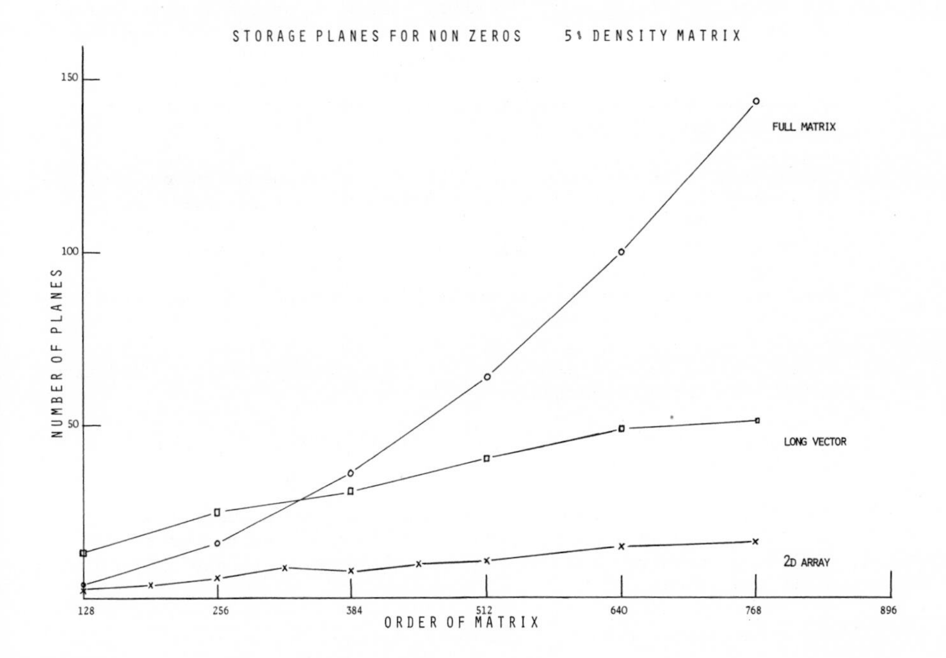
STORAGE PLANES FOR NON ZEROS 5% DENSITY MATRIX
NUMBER OF PLANES
150
100
50
FULL MATRIX
LONG VECTOR
2D ARRAY
128
256
384
512
640
768
896
ORDER OF MATRIX

Fig. 5

7. FULL MATRIX VECTOR MULTIPLICATION

If sufficient storage is available then the matrix vector multiplication can be carried out by partitioning the matrix and vector into 64 x 64 blocks and by performing the multiplications in parallel treating all coefficients as non zero. This process can be organised with very little in the way of overheads and consequently becomes competitive with the sparse matrix algorithms when the density of non zeros is as low as about 5% of the total elements. A test to avoid multiplication of null blocks would be included.

8. COMPARISON OF THE TWO ALGORITHMS

(1) Both algorithms require far less storage than full matrix multiplication.

(2) Fast execution is possible for sparse (<5%) matrices.

(3) For sparsity of >>5%, full matrix multiplication is faster (if the storage is available) than either algorithm.

(4) "Long Vector" can only deal with N≤4096; "2D Array" is not restricted in this way. "2D Array" is only restricted by the size of the DAP store since it deals with block 64 x 64 partitioning.

(5) Storage requirements for the two algorithms are problem dependent. If the non zeros happen to be numerous in one or more rows then the storage planes required by "Long Vector" will be high. The block partitioning of "2D Array" tends to spread the non zeros of the unstructured matrix more uniformly and "2D Array" required about one third of the planes used by "Long Vector" in the examples tested (see Figs. 4, 5). With a structured Matrix there will be examples in which the storage plane requirement of the two algorithms is reversed.

(6) "Long Vector" is restricted to less than 50 elements in any row; "2D Array" is not restricted in the same way but is unable to store very many blocks if the non zeros occur in the same relative position in most of the blocks.

(7) Both can cope with rectangular matrices and store indices using only short length variables.

(8) Vector matrix multiplication can be performed by a slight variation in the "2D Array" algorithm.

(9) Sparsity in vector $\underline{x}$ can be taken advantage of in "2D Array" but not in "Long Vector".

(10) Repositioning of the non zeros amongst the processing elements can lead, in each algorithm, to a reduction in the number of storage planes required and hence an improvement in the number of plane operations.

(11) The final vector result is produced unpacked by "2D Array".

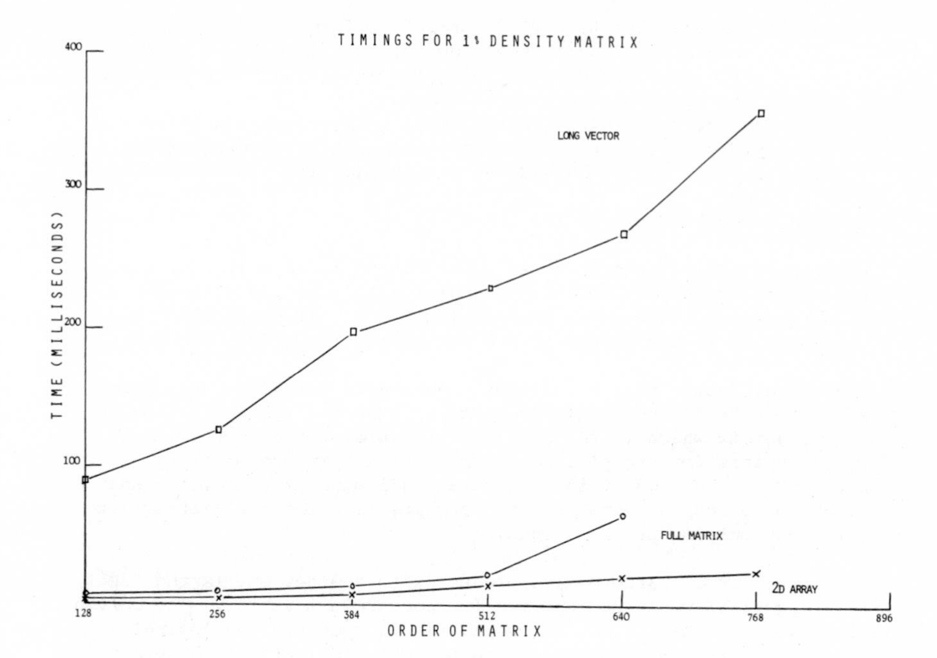
TIMINGS FOR 1% DENSITY MATRIX
TIME (MILLISECONDS)
400
300
200
100
LONG VECTOR
FULL MATRIX
2D ARRAY
128
256
384
512
640
768
896
ORDER OF MATRIX

Fig. 6

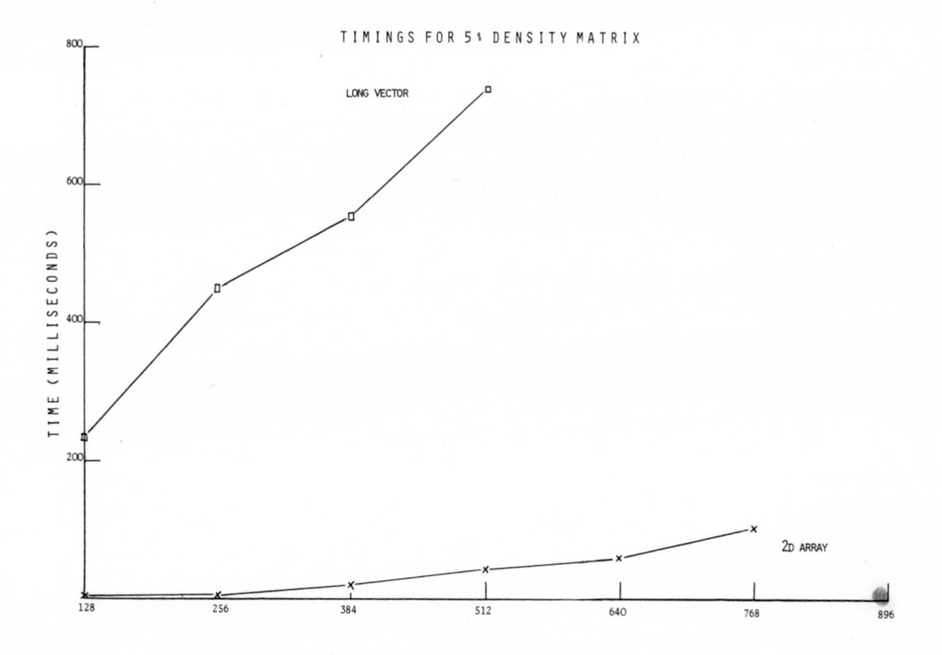
TIMINGS FOR 5% DENSITY MATRIX
TIME (MILLISECONDS)
800
600
400
200
LONG VECTOR
2D ARRAY
128
256
384
512
640
768
896

Fig. 7

(12) In all the examples tested "2D Array" performed significantly faster than "Long Vector" see Figs. 6, 7. It could be speeded up by using micro coded routines which is already done in "Long Vector" to optimise the permute stage.
Although the full matrix multiplication algorithm can be carried out rapidly on comparatively small problems the storage requirements rule it out of consideration for practical sparse problems.

9. FURTHER IMPROVEMENTS TO "2D ARRAY"

By separating out the diagonal and treating it as a long vector a speed up will generally be obtained particularly if the diagonal is full as would be the case in many applications. It would also lead to a reduction in the number of storage planes required.

The final vector result is readily available for algorithms which need to operate on the resulting vector. Repeated multiplication by the same matrix which is the essential feature of the power method for eigenvalues for example, can be performed very economically if the vector in the product is treated as full since the logical masks created to correspond with the non zero positions of the coefficient matrix need only be produced once.

To carry out a sparse matrix product with another matrix the "2D Array" algorithm would be repeated for each column vector. The column vector can be packed if the second matrix is sparse. The logical arrays would then have to be constructed repeatedly for each column vector but the packing of the column vector would reduce very considerably the number of multiplication calls which would otherwise have to be made.

REFERENCES

Morjaria, M. and Makinson, G.T. (1981) Operations with Sparse Matrices on the ICL DAP, Mathematical Institute, University of Kent at Canterbury, Canterbury, Kent.

Parkinson, D. (1981) Sparse Matrix Vector Multiplication on the DAP, DAP Support Unit, Queen Mary College, London.

BAND MATRICES ON THE DAP

L.M. Delves and A.S. Samba

(Department of Statistics and Computational Mathematics, University of Liverpool)

and

J.A. Hendry

(Computer Centre, University of Birmingham)

1. INTRODUCTION

Our major interest in the DAP is as a vehicle for solving elliptic partial differential equations; we are currently mounting the PDE package GEM2 (Phillips, Delves and O'Neill (1982)) on the ICL 2980 at QMC and studying alternative ways to transfer this, in whole or in part, to the DAP.

GEM2 implements the Global Element Method (Delves and Hall (1979) and Delves and Phillips (1980)), a variant of the Finite Element Method using a small number of high (variable) order elements. The current serial implementation, designed to reduce the serial operations count, makes heavy use of FFT techniques for setting up the defining equations, and of block matrix manipulations during the solution phase. The size of the blocks is related to the order of approximation in a single element, an approximation of order 8 (a modest order in this context) yielding for the two-dimensional problems handled by GEM2, blocks of size 64 x 64; this obvious high degree of inherent parallelism makes the DAP a very interesting target machine for the method. Of course, it is not self-evident that the detailed algorithms developed in a serial environment are those which will behave best on the DAP; but, with a large package, it is inevitable that the first stage of development will be to transfer parts of the existing code to the DAP, and hence preserve the existing structure.

We have been looking first at the solution stage. With M elements in q dimensions, with an approximation of order N in each element, this involves the solution of an $MN^q \times MN^q$ square system of equations, with M x M blocks of size $N^q \times N^q$. Off-diagonal blocks are empty (where two elements have no side in common) or full but of low rank (where elements touch), and the current solution technique (Hendry, Delves and Mohamed (1981)) uses the low rank nature of the blocks to decouple the problem, leaving a sequence of M full $N^q \times N^q$ matrix problems to solve, involving the diagonal blocks. These blocks are also heavily structured, and an iterative technique for solving these is described in (Delves and Phillips (1980)) which requires the direct solution of a band system of relatively small bandwidth, with a sequence of right hand sides. We are therefore led to look at the sub-problem: how should band systems be solved on the DAP? We report progress on this problem here; results obtained with a model one-dimensional GEM program on the DAP, using the techniques described here, are reported elsewhere in these proceedings (Hendry and Delves (1983)).

2. CURRENT ALGORITHMS

The simplest band matrix is a tridiagonal matrix (semi-bandwidth $\mu = 1$); for such systems the cyclic reduction algorithm can be implemented in parallel form, and a routine for this exists in the DAP subroutine library (Whiteway (1979)). By partitioning into blocks of size $\mu \times \mu$, any band matrix of semi-bandwidth μ can be written as a block tridiagonal matrix, and cyclic reductions performed on the block system. A routine, due to Hellier (1981), based on this idea is also in the DAP subroutine sub-library; it treats only quin-diagonal ($\mu = 2$) systems.

3. PARALLEL CYCLIC REDUCTION

We investigate first the implementation of parallel block cyclic reduction, for arbitrary bandwidth μ. In the GEM context, a number of such systems must be solved, with different matrices; our implementation provides for parallel solution of as many systems as can be packed into a DAP matrix, but is is simpler to first describe the algorithm for a single 64 x 64 system. For illustrative purposes we set $\mu = 8$; then the matrix takes the block form

$$\begin{matrix}
a_1 & b_1 & \cdot & \cdot & \cdot & \cdot & \cdot & \cdot \\
c_2 & a_2 & b_2 & \cdot & \cdot & \cdot & \cdot & \cdot \\
\cdot & c_3 & a_3 & b_3 & \cdot & \cdot & \cdot & \cdot \\
\cdot & \cdot & c_4 & a_4 & b_4 & \cdot & \cdot & \cdot \\
\cdot & \cdot & \cdot & c_5 & a_5 & b_5 & \cdot & \cdot \\
\cdot & \cdot & \cdot & \cdot & c_6 & a_6 & b_6 & \cdot \\
\cdot & \cdot & \cdot & \cdot & \cdot & c_7 & a_7 & b_7 \\
\cdot & \cdot & \cdot & \cdot & \cdot & \cdot & c_8 & a_8
\end{matrix}$$

The first stage of the algorithm has two sub-stages:

(i) Reduce the diagonal blocks to diagonal form using a Gauss-Jordan routine. This can be done in parallel on all blocks, and is equivalent to pre-multiplying the ith row of (1) by a_i^{-1} (assumed non-singular); the matrix then has the form

$$\begin{array}{cccccccc}
I & b_1' & . & . & . & . & . & . \\
c_2' & I & b_2' & . & . & . & . & . \\
. & c_3' & I & b_3' & . & . & . & . \\
. & . & c_4' & I & b_4' & . & . & . \\
. & . & . & c_5' & I & b_5' & . & . \\
. & . & . & . & c_6' & I & b_6' & . \\
. & . & . & . & . & c_7' & I & b_7' \\
. & . & . & . & . & . & c_8' & I
\end{array}$$

(ii) Eliminate b_i' from row i by subtracting b_i' * row (i+1), i = 1, 2,...,7 and eliminate c_1' from row i by subtracting c_i' * row (i-1), i = 2, 3,...,8. These operations can also be carried out in parallel for each i; the matrix then takes the form

$$\begin{array}{cccccccc}
x & . & x & . & . & . & . & . \\
. & x & . & x & . & . & . & . \\
x & . & x & . & x & . & . & . \\
. & x & . & x & . & x & . & . \\
. & . & x & . & x & . & x & . \\
. & . & . & x & . & x & . & x \\
. & . & . & . & x & . & x & . \\
. & . & . & . & . & x & . & x
\end{array}$$

where x stands for a non-null block and "." for a null block. We see that the super and sub-diagonals have been "swept out" one step. If we repeat the process of (i) diagonalising diagonal blocks; (ii) eliminating the super and sub-diagonals we find that at the end of stage 2 the matrix has the form

$$
\begin{array}{cccccccc}
x & \cdot & \cdot & \cdot & x & \cdot & \cdot & \cdot \\
\cdot & x & \cdot & \cdot & \cdot & x & \cdot & \cdot \\
\cdot & \cdot & x & \cdot & \cdot & \cdot & x & \cdot \\
\cdot & \cdot & \cdot & x & \cdot & \cdot & \cdot & x \\
x & \cdot & \cdot & \cdot & x & \cdot & \cdot & \cdot \\
\cdot & x & \cdot & \cdot & \cdot & x & \cdot & \cdot \\
\cdot & \cdot & x & \cdot & \cdot & \cdot & x & \cdot \\
\cdot & \cdot & \cdot & x & \cdot & \cdot & \cdot & x
\end{array}
$$

The super and sub diagonals have been swept out twice as far in Stage 2 as in Stage 1; Stage 3 will sweep them at twice as far again, and so on. Hence, if the block matrix is of size N x N, $[\log_2 N]$ stages sweep the super and sub-diagonals completely away, leaving only the diagonal blocks; the matrix takes diagonal form and the system can be solved trivially.

We have implemented this process on the DAP, allowing for a number of systems to be stored and treated simultaneously. The diagonal matrices a_i, $i = 1,\ldots, N$ are stored in columns of a DAP matrix. For an underlying 64 x 64 matrix of semi-bandwidth μ, the a_i will occupy μ columns, and we store the a_i for each of the systems in adjacent columns. The b_i and c_i are similarly stored in columns, with a zero block adjoined to pad out the columns ($b_1 = 0$, $c_N = 0$), as are the right hand sides; the program assumes there are μ right hand sides for each system, since this is appropriate in the GEM context. It is therefore possible to store $S = [64/\mu]$ systems within four DAP matrices, and to solve these in parallel. Table 1 shows the timings (per system, with μ RHS) obtained in this way using REAL*8 arithmetic. The overall time spent is split roughly evenly in this algorithm between reducing the diagonal blocks using a full Gauss-Jordan technique, and multiplying blocks together; these times are shown separately in Table 1. For comparison, the time to solve a single 64 x 64 <u>full system</u>, but with only a single RHS, using Gauss-Jordan, is 175 msec; we see that, for 64 x 64 band matrices, this technique is at least faster than treating the matrix as full for semi-bandwidth $\mu < 12$.

Table 1

Solution (in a REAL*8) of block tri-diagonal systems using Block Parallel Cyclic Reduction - blocks of size μ by μ. Solution in Parallel of a number of linear independent tri-diagonal systems (of size 64 x 64) with μ independent RHS.

μ	Total Solution Time (msecs)	Multiplication Time (msecs)	Gauss-Jordan Time (msecs)	No. of Linear Systems	Time per Linear System
4	409	216	193	16	25
6	595	303	292	10	60
8	581	283	298	8	73
10	747	350	397	6	125
12	894	419	475	5	179
16	809	352	457	4	202
20	1090	450	640	3	363
24	1327	550	777	2	663
28	1570	660	910	2	885
32	961	317	644	2	480

4. PARALLEL REDUCTION OF AUGMENTED TRIANGLES (PRAT)

Block Parallel Cyclic Reduction has three disadvantages viewed purely as a technique for solving band matrices:

(1) The partitioning of a band matrix in tri-diagonal form yields super and sub-diagonal blocks which are triangular; but the technique takes no advantage of this.

(2) While it is easy to process efficiently several RHS simultaneously, it is not possible to re-enter the routine with further right hand sides without repeating the whole reduction process.

(3) The diagonal blocks are assumed non-singular throughout, and no pivoting between blocks is possible, although pivoting within blocks can be achieved. We have therefore looked for other schemes; and discuss one here (the PRAT technique) for which we comment on points (1) to (3) as follows:

Point (1) : PRAT takes full advantage of the form of the band matrix

Point (2) : Not yet studied

Point (3) : This assumption is replaced by a much hairier one - see below

The PRAT technique also turns out to be very fast, even for a single system.

We consider the band system

$$W\,\underline{u} = \underline{v} \quad ; \quad w_{ij} = 0.0 \text{ if } |i-j| > \mu \qquad 1 \le i,j \le N. \tag{1}$$

With ℓ defined as: $\ell = N/2\mu$; $O(\log_2 \ell)$ parallel major steps of a dynamic partitioning are required to solve this band system, as follows.

We first partition W into 2ℓ- conformable submatrices A_i, B_i, $i=1,2,..,\ell(=N/2\mu)$, as shown below

$$W \equiv \begin{bmatrix} A_1 & B_1 & & & \\ & A_2 & B_2 & & \\ & & A_3 & B_3 & \\ & & & & \\ & & & A_\ell & B_\ell \end{bmatrix}$$

The submatrix A_1 is a <u>rectangular</u> block matrix of dimensions $2\mu \times \mu$; A_i, $i=2,..,\ell$ are <u>upper triangular</u> with dimensions $2\mu \times 2\mu$ respectively.

The submatrices B_i, $i=1,2,...,\ell-1$ are lower triangular with dimensions $2\mu \times 2\mu$. The final submatrix is also assumed to be a <u>lower</u> triangular matrix of dimensions $2\mu \times 2\mu$; in order to achieve this lower triangular structure we pad B_ℓ with μ additional columns, putting fictitious non-zero (unit) elements along the lower half of its principal diagonal and zeros elsewhere.

A schematic representation of the partitioning scheme described above is given in Fig. 1.

The solution vector $\underline{u}$ is partitioned conformably into ℓ $2\mu^{th}$ order subvectors $\underline{x}_i^T$ $i=2,3,..,\ell+1$ and a μth order vector $\underline{x}_1^T$ respectively:-

$$x_i^T = \{u_k; k = \max[1, (2i-3)\mu+1], \dots (2i-1)\mu\} \tag{2}$$

with

$$\underline{u}^T = [\underline{x}_1^T, \underline{x}_2^T, \dots \underline{x}_\ell^T, \underline{x}_{\ell+1}^T]. \tag{3}$$

We accommodate the fictitious elements introduced in B_ℓ by assuming that the components u_k of the vector $\underline{u}^T$ are zero if $k>N$.

Finally, the partitioning is performed on the vector $\underline{v}$ appearing on the right hand side of (1).

$\underline{v}^T$ is decomposed into ℓ $2\mu^{th}$ order subvectors

$$\underline{y}_i^T \quad i=1,2,\dots \ell: \tag{4}$$

$$y_i = \{\underline{v}_k \; ; \; k=1 + 2(i-1)\mu, \dots, 2i\mu\}$$

with

$$\underline{v}^T = [\underline{y}_1^T, \underline{y}_2^T, \dots \underline{y}_\ell^T]. \tag{5}$$

With this partitioning, (1) is equivalent to the system of equations

$$A_i\underline{x}_i + B_i\underline{x}_{i+1} = \underline{y}_i. \tag{6}$$

Consider now the jth equation of the above system:

$$A_j\underline{x}_j + B_j\underline{x}_{j+1} = \underline{y}_j.$$

Normalising with respect to B_j and writing the normalised coefficients with zero superscript gives

$$A_j^{(0)} \underline{x}_j + \underline{x}_{j+1} = \underline{y}_j^{(0)}. \tag{7}$$

The corresponding equation for the $(j-1)^{th}$ equation is

$$A_{j-1}^{(0)} \underline{x}_{j-1} + \underline{x}_j = \underline{y}_{j-1}^{(0)}. \tag{8}$$

Substituting in (7) for $\underline{x}_j$ from (8) gives

$$-A_j^{(0)} A_{j-1}^{(0)} \underline{x}_{j-1} + \underline{x}_{j+1} = \underline{y}_j^{(0)} - A_j^{(0)} \underline{y}_{j-1}^{(0)}. \tag{9}$$

The corresponding equation for the $(j-2)^{th}$ equation is now

$$-A_{j-2}^{(0)} A_{j-3}^{(0)} \underline{x}_{j-3} + \underline{x}_{j-1} = \underline{y}_{j-2}^{(0)} - A_{j-2}^{(0)} \underline{y}_{j-3}^{(0)} . \qquad (10)$$

Substituting now in (8) for $\underline{x}_{j-1}$ from (10) gives

$$A_j^{(2)} \underline{x}_{j-3} + \underline{x}_{j+1} = \underline{y}_j^{(2)} \qquad (11)$$

where

$$\begin{aligned} A_j^{(0)} &= B_j^{-1} A_j \\ A_j^{(1)} &= - A_j^{(0)} A_{j-1}^{(0)} \\ A_j^{(2)} &= A_j^{(0)} A_{j-1}^{(0)} A_{j-2}^{(0)} A_{j-3}^{(0)} \\ y^{(0)} &= B_j^{-1} \underline{y}_j \end{aligned} \qquad (12)$$

and

$$\underline{y}_{(j)}^{(1)} = \underline{y}_j^{(0)} - A_j^{(0)} \underline{y}_{j-1}^{(0)}$$

$$y_j^{(2)} = \underline{y}_j^{(0)} - A_j^{(0)} \underline{y}_{j-1}^{(0)} + A_j^{(0)} A_{j-1}^{(0)} (\underline{y}_{j-2}^{(0)} - A_{j-2}^{(0)} \underline{y}_{j-3}^{(0)}).$$

Equation (11) can be expanded in terms of $\underline{x}_{j-3}$ to get a relation in $\underline{x}_{j-7}$ and $\underline{x}_{j+1}$. The substitution continues, the kth stage related to the $(k-1)^{th}$ stage by

$$A_j^{(k)} = - A_j^{(k-1)} A_{j-k}^{(k-1)} \qquad (13)$$

and

$$\underline{y}_j^{(k)} = \underline{y}_j^{(k-1)} - A_j^{(k-1)} \underline{y}_{j-k}^{(k-1)} .$$

These stages are illustrated in Figs. 1 - 5 for a 32 x 32 matrix of semi-bandwidth $\mu = 2$.

Following the nth stage of substitution we get

$$A_i^{(n)} \underline{x}_1 + \underline{x}_{i+1} = \underline{y}_i^{(n)} \qquad i=1,2,\ldots \ell \qquad (14)$$

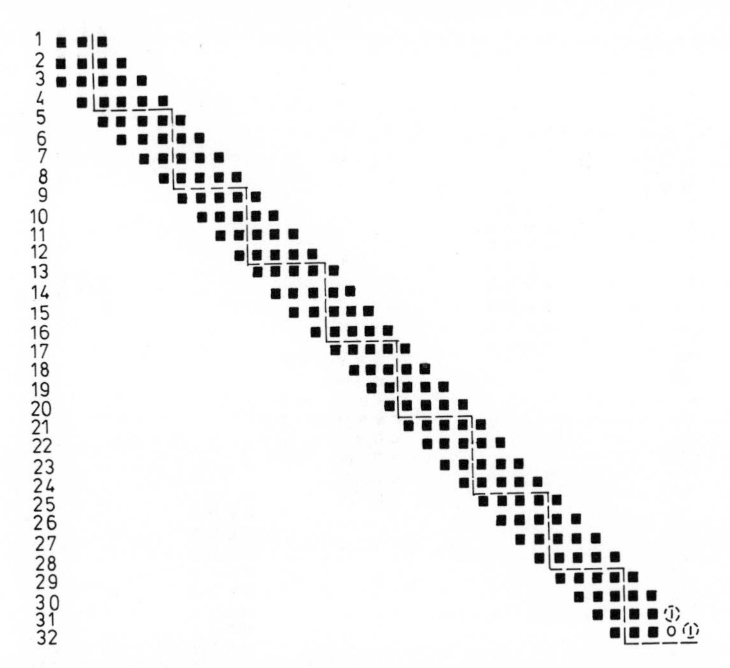

Fig. 1 The PRAT partitioning for n=32, μ=2. Note the initial non-triangular block, and the padding of the final block to triangular form

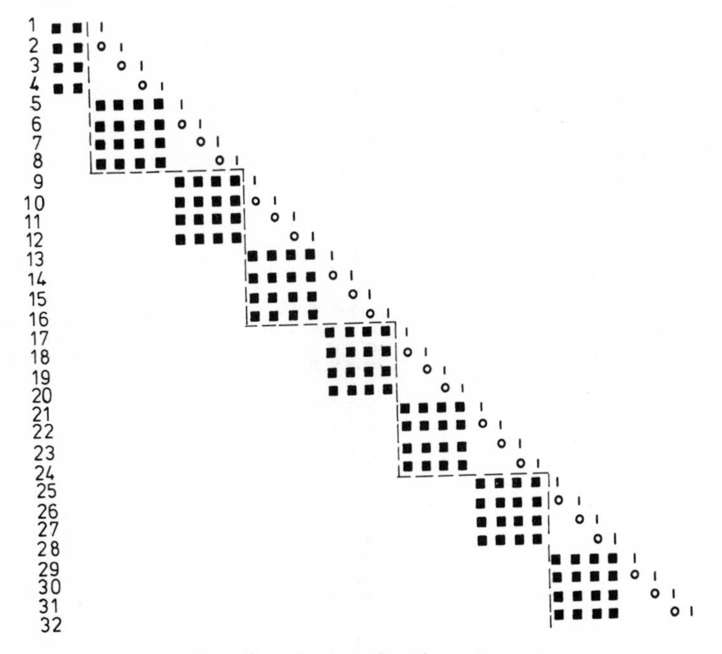

Fig. 2 First reduction stage

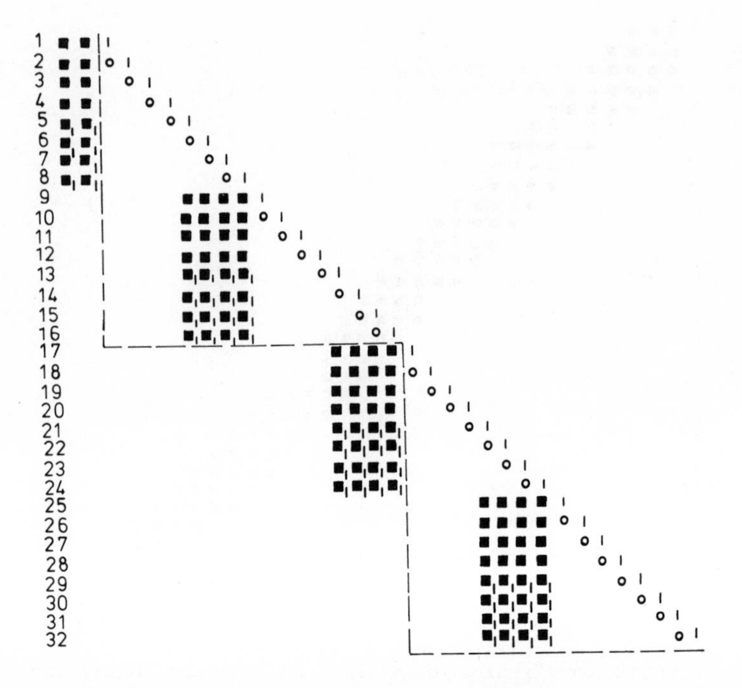

Fig. 3 Second reduction stage

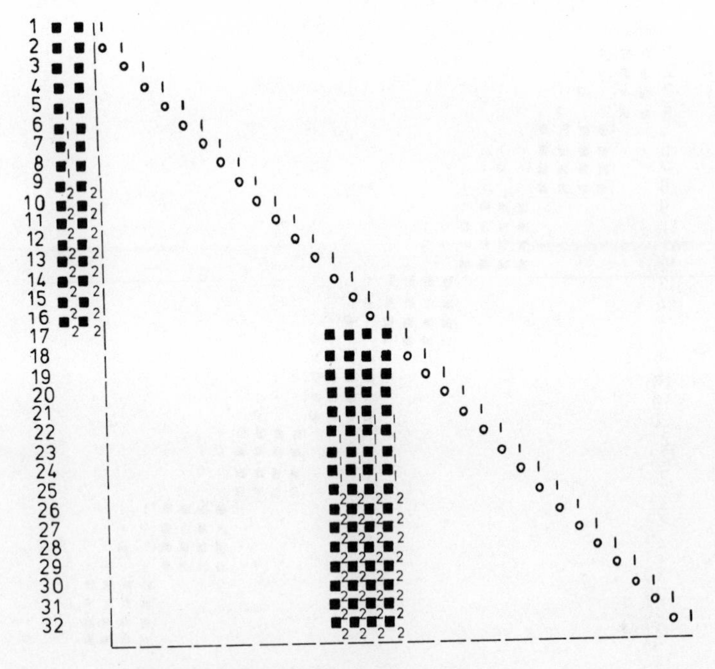

Fig. 4 Third reduction stage

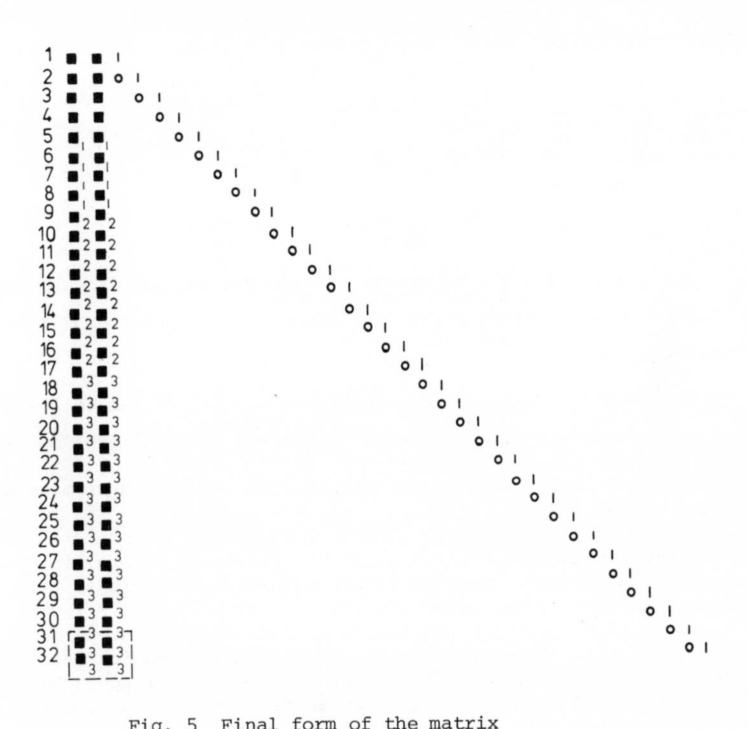

Fig. 5 Final form of the matrix

The superscript n is given by $n = [\log_2 \ell]$,

Consider now the $(\ell + 1)^{th}$ subvector $\underline{x}_{\ell+1}$:

Using (2) the components of $\underline{x}_{\ell+1}$ are given by

$$\underline{x}^T_{\ell+1} = [u_k \; / \; k = 1+N-\mu \; ,\ldots \; \ldots ,N+\mu]$$

$$= [u_{N+1-\mu}, \; u_{N+2-\mu}, \; \ldots \; u_N, \; 0,0,\ldots\ldots,0] \tag{15}$$

where the trailing zeros represent the μ fictitious variables, set here to zero.

Consider the ℓth system of equations in (14):

$$A_\ell^{(n)} \underline{x}_1 + \underline{x}_{\ell+1} = \underline{y}_\ell^{(n)}$$

Substituting in the above for $\underline{x}_{\ell+1}$ from (15) gives

$$\sum_{j=1}^{\mu} A_{\ell}^{(n)}(i,j)\ \underline{x}_1^{(j)} = \underline{y}_{\ell}^{(n)}(i),\ i = \mu + 1,\ldots,2\mu \qquad (16)$$

Equation (16) is a $\mu \times \mu$ subsystem of linear equations for the unknown μ^{th} order vector $\underline{x}_1$. We solve (16) for $\underline{x}_1$ using a standard Gauss-Jordan routine.

With $\underline{x}_1$ known the remaining subvectors $\underline{x}_j$, $j=2,..,\ell+1$ follow from (14), and hence the solution vector $\underline{u}$ for the original problem (1).

Notice that for a tridiagonal system the $\mu \times \mu$ subsystem is simply a 1 x 1 system.

Timings for an implementation of this reduction scheme, in REAL*4, are given in table 2, compared with alternative reductions: tri-diagonal (point) cyclic reduction for $\mu=1$ (Whiteway (1979)); Hellier's quin-diagonal routine for $\mu=2$ (Hellier (1981)) and Gauss-Jordan (full matrix) for $\mu>2$. These timings are encouraging; even for $\mu=1$, the technique appears slightly faster than the best existing technique, while for $\mu>1$ it appears about twice as fast as previous routines. Comparison with table 1 (block cyclic reduction) is less direct. Allowing for the difference in arithmetic precision, the measured times are about the same; however, the block cyclic routine solves a number of systems in parallel, and the times given are the time per system. Extending the PRAT routine in the same way would certainly reduce the time per system significantly.

The PRAT technique has, however, the built-in assumption that the triangular blocks B_i are non-singular; this assumption will be violated (for example) if any element on the topmost super-diagonal should be zero, and we can expect trouble if such elements are small. We are currently investigating several methods of avoiding this restriction, and those interested are referred to (Samba (1982)).

5. MATRIX MULTIPLICATION

As a sub-task in both the Cyclic Reduction and PRAT algorithms, we need to multiply sets of small matrices together. The "standard" method for doing this on the DAP is prescribed by Gostick (1979) or McKeown (1981). We have had two attempts at improving on their timings; and since the problem is of common occurrence, we report these here.

Table 2

Solution of a simple Band Matrices system using the PRAT algorithm
SYSTEM SIZE 64 x 64; REAL*4 ARITHMETIC

BANDWIDTH	PRAT	COMPETING ROUTINE
μ	(BS)	
1	5.06 ms	5.88
5	30.00 ms	61 ms
6	32.90 ms	73.6 ms (Gauss Jordan)
7	35.90 ms	73.6 ms
8	38.90 ms	73.6 ms

Version 1

We assume, as is common, that several matrix multiplications of the form C = A*B, where A, B and C are $\mu \times \mu$, are to be carried out. We pack as many of these systems as we can into a DAP matrix (so that for $\mu = 8$ we can fit in 64 independent matrices), and implement an outer product algorithm, generalised to handle many blocks at once. The code extends the standard outer product algorithm to many blocks at once by replacing the DAP matrix broadcasting functions (MATC, MATR) by equivalent cascade shifts over relevant columns/rows of a DAP matrix. Timings in REAL*8 and REAL*4 are given in table 3, where they are compared with those obtained from an implementation of the algorithm of Gostick. We see that, for $\mu \leqslant 8$, this method is faster per system than that of Gostick, but for larger values of μ, Gostick's method is faster.

Table 3

Matrix Multiplication Timings : Version 1
Timings given are in msecs, for the number of matrix multiplications shown

		REAL*8		REAL*4	
μ	No. of independent products	Gostick	"Version 1"	Gostick	"Version 1"
2	1024	6.2	3.3	2.7	1.3
4	256	9.8	7.3	4.1	2.9
6	100	13.6	12.5	5.5	5.2
8	64	17.6	16.7	7.1	7.1
10	36	21.9	24.7	8.8	10.8
12	25	26.4	23.9	10.6	13.1
14	16	31.9	35.2	12.5	15.4
16	16	35.8	40.2	14.5	17.8
18	9	40.9	55.6	16.6	15.4
20	9	46.3	62.7	18.8	28.6

Version 2

In this version we try to improve on the timings obtained in Version 1, by the use of partitioning techniques.

Consider the problem of evaluating the product

$$A \times B$$

where A has dimension $(N_1 \times N_2)$

B " " $(N_2 \times N_3)$

and suppose $N_2 \leqslant N_1 \leqslant 32$, and let $\ell = N_2/2$.

We partition A, B as $A = (A_1 \; A_2)$ and $B = \begin{pmatrix} B_1 \\ B_2 \end{pmatrix}$

where A_1, A_2 have dimensions $(N_1 \times \ell)$

B_1, B_2 have dimensions $(\ell \times N_3)$

then $C = A \times B = A_1 \times B_1 + A_2 \times B_2$.

On the DAP, we store the partitioned matrix (A) (B) by columns:

$$A \leftarrow \begin{bmatrix} A_1 \\ A_2 \end{bmatrix} \qquad B \leftarrow \begin{bmatrix} B_1 \\ B_2 \end{bmatrix}$$

and compute C as follows:

```
         C = 0.0
For      J = 1, 2 .. UNTIL ℓ do.
         C = C + MATC(A(,J))*MERGE(MATR(B(J+ℓ,)), MATR(B(J,)), MASK)
```

where 'MASK' is a logical matrix defined as

```
         MASK =  ALTR(ℓ)
```

Then compute

$$C(i,j) \leftarrow C(i,j) + C(i+\ell,j,) \; ; \; 1 \leqslant i \leqslant \ell$$

$$j = 1(1)\, N_3$$

i.e. $C = C + \text{SHNP}\ (C,\ \ell)$

NOTICE that the above technique requires

(i) $(\ell + 1)$ additions

(ii) ℓ multiplications

Timings for REAL*4 arithmetic for this strategy, are given in Table 4, for two subversions:

(1) Subversion 2.1

A simple matrix multiply. These times are compared with the best given in McKeown (1981).

(2) Subversion 2.2

Multiple simultaneous matrix multiplies. These times should be compared with those in table 3.

In Subversion 2.1, a modest but significant improvement is shown; it is possible that further improvements for $\mu > 2$ would follow given additional partitioning of the matrices. Subversion 2.2 beats McKeown (1981), but gives results slightly slower than those of Version 1.

Table 4

Matrix Multiplication : Version 2.2
Timings for REAL*4 arithmetic, in msec

μ	Number of systems	Version 2	McKeown (1981)
2	512	0.86	-
2	1024	1.59	-
4	128	1.58	-
4	256	3.02	-
8	1	3.46	-
16	1	6.55	7.08
32	1	12.76	14.2

ACKNOWLEDGEMENTS

The authors are grateful to the SERC for support under grant GR/B/24332 to the Universities of Birmingham and Liverpool.

REFERENCES

Phillips, C., Delves, L.M. and O'Neill, T. (1982) GEM2: A Program for the Solution of Elliptic PDE's, submitted to TOMS.

Delves, L.M. and Hall, C.A. (1979) An Implicit Matching Principle for Global Element Calculations, *J. Inst. Math. Applics.*, **23**, pp. 223-234.

Delves, L.M. and Phillips, C. (1980) A Fast Implementation of the Global Element Method, *J. Inst. Math. Applics.*, **25**, pp. 177-197.

Hendry, J.A., Delves, L.M. and Mohamed, J.L. (1982) Iterative Solutions of the Global Element Equations, in MAFELAP1981 (J.R. Whiteman, ed.), Academic Press, pp. 77-84.

Hendry, J.A. and Delves, L.M. (1982) The Global Element Method on the ICL DAP, these proceedings.

Whiteway, J. (1979) A Parallel Algorithm for Solving Tri-diagonal Systems, DAP Newsletter 3, Queen Mary College.

Hellier, R. (1981) A Quindiagonal Solver for the DAP, University of Canterbury at Kent.

Samba, A.S. (1982) Ph.D. Thesis, University of Liverpool.

Gostick, R.W. (1979) Software and Algorithms for the DAP, *ICL Techn. J.*, **1**, pp. 116-135.

McKeown, J.J. (1981) Multiplication of non-standard Matrices on the DAP, DAP Newsletter3, Queen Mary College.

GEM CALCULATIONS ON THE DAP

J.A. Hendry

(Centre for Computing and Computer Science, University of Birmingham, Birmingham)

L.M. Delves

(Department of Statistics and Computational Mathematics, University of Liverpool, Liverpool)

ABSTRACT

The results of a pilot implementation of the Global Element Method (GEM) on the DAP are presented. Emphasis is concentrated on the solution of the linear equations arising from the GEM.

1. INTRODUCTION

The Global Element Method (GEM, see Delves and Hall (1979)) for the solution of elliptic partial differential equations can be viewed as a variant of the Finite Element family using large, variable order, non-conforming elements with special provision for curved boundaries, inter-element continuity, a variety of boundary conditions and singularities. The basic philosophy is to use a fixed number M of elements, and to improve the solution by increasing the degree (N-1) of the approximation in each element. In practice, very rapid convergence can be obtained as N increases (see Hendry, Delves and Phillips (1978) and Hendry (1980)).

A GEM calculation has two distinct phases:

(i) the assembly, by a numerical technique, of the appropriate matrix

(ii) the solution of the resulting sets of linear equations.

The attractiveness of any numerical method depends both on its accuracy and costs, and special techniques have been developed (see Delves and Phillips (1980)) to reduce the operation count (on a serial machine) for both phases from the somewhat uneconomic costs that a straightforward approach would suggest.

In this paper, the results of a pilot study of GEM calculations on the ICL DAP are reported. For simplicity, we restrict attention to a one-dimensional problem. In section 2 a brief outline of the important features of the GEM is given, while section 3 discusses in more detail approaches to the solution of the GEM equations. Section 4 contains a comparison of the results for the DAP and a serial machine (CDC 7600), and finally section 5 contains some conclusions. The results presented here only apply to the solution phase. This restriction is not only for brevity, but also reflects the fact that the solution phase tends to dominate the overall costs of a GEM calculation.

The technique used for the solution phase is based on a low-rank modification scheme (see Delves and Hall (1979) for more details), which, for a one-dimensional problem, replaces the solution of a single MN x MN linear system by the solution of M independent N x N linear systems, followed by the subsequent assembly and solution of an auxiliary 4M x 4M linear system. The motivation for this pilot study stems from noting that the parallelism of the DAP can be exploited by performing simultaneously the solution of the M independent N x N linear systems.

2. DESCRIPTION OF GEM

We consider here a one-dimensional equation over the range $a \leq y \leq b$. In the GEM, the range (a,b) is subdivided into M elements (y_i, y_{i+1}) such that,

$$a = y_0 < y_1 < y_2 ----- y_{m-1} < y_m = b$$

and after transformation of the variables, the general form of the equation in element i is

$$\left[-\frac{d}{dx}\left(A(x)\frac{d}{dx}\right) + B(x)\right]f(x) = g(x) \quad -1 \leq x \leq 1 \tag{2.1}$$

In element i, a solution of the form

$$U^{(i)}(x) = \sum_{k=1}^{N} a_k^{(i)} h_k^{(i)}(x)$$

is sought where the $h_k^{(i)}(x)$ are known basis functions ($h_k^{(i)}(x)$ is a polynomial of degree k-1, see later) and $a_k^{(i)}$ are the unknown parameters.

The GEM then replaces the differential equation by the coupled linear equations

$$(\underline{\underline{V}}_{ii} + \underline{\underline{S}}_{ii})\underline{a}^{(i)} = \underline{g}^{(i)} - \sum_k \underline{\underline{S}}_{ik}\, \underline{a}^{(k)} \qquad i = 1,....,M \tag{2.2}$$

where the summation in k is over the elements adjacent to element i (in this case, i-1 and i+1). In eq. (2.2), the vector $\underline{a}^{(i)}$ is the vector of unknown coefficients $a_k^{(i)}$ in the ith element, while the elements of the matrix $\underline{\underline{V}}_{ii}$ are obtained from a numerical integration over element i. The elements of $\underline{\underline{S}}_{ij}$ reflect the coupling of an element with itself or its neighbours, and do not require numerical integration.

As an illustration of the numerical techniques involved, the B(x) term in (2.1) contributes to the (p,q) element of $\underline{\underline{V}}_{ii}$ as

$$\int_{-1}^{+1} h_p^{(i)}(x)\ B(x)\ h_q^{(i)}(x)\,dx \tag{2.3}$$

Making the specific choice for $h_p^{(i)}(x)$, (see Delves and Phillips (1980))

$$h_p^{(i)}(x) = \begin{cases} T_0(x) & p = 1 \\ T_1(x) & p = 2 \\ (1-x^2)T_{p-3}(x) & p = 3,\ldots\ldots \end{cases} \tag{2.4}$$

where $T_n(x)$ is a Chebyshev polynomial, allows the integration of (2.3) to be performed by FFT, rather than by conventional numerical quadrature techniques, with a consequent saving in costs.

Each $\underline{\underline{S}}_{ij}$ can be written in the form

$$\underline{\underline{S}}_{ij} = \sum_{\ell=1}^{4} \gamma_\ell^{(i,j)} \underline{\alpha}_\ell^{(i)} \underline{\beta}_\ell^{(j)T} \tag{2.5}$$

where $\underline{\alpha}_\ell^{(i)}$, $\underline{\beta}_\ell^{(j)}$ are N vectors (essentially the basis functions and their derivatives evaluated at the end points of elements i,j) and $\gamma_\ell^{(i,j)}$ is a scalar. With the form (2.4) for $h_p^{(i)}(x)$, the structure of $\underline{\underline{S}}_{ij}$ is particularly simple.

Further details of the method and its implementation (together with the extension to higher dimension problems) can be found in Delves and Phillips (1980).

The results presented later in section 4 refer to the particular equation

$$-\frac{d^2F}{dy^2} + F = \exp\,(S)\left[1-(3\pi C)^2 + (3\pi)^2 S\right] \qquad 0 \leqslant y \leqslant M \tag{2.6}$$

with boundary conditions

$$F(0) = F(M) = 1$$

where

$$S = \sin\,(3\pi y) \qquad C = \cos\,(3\pi y)$$

The exact solution is

$$F(y) = \exp(S)$$

For convenience equal sized elements were used, the A(x) etc. of eq. (2.1) being obtained in each element i by appropriate linear transformations on the terms of (2.6).

3. SOLUTION OF GEM EQUATIONS

3.1 Decoupling

The structure (2.5) of the $\underline{\underline{S}}_{ij}$ permits a low rank modification technique to be applied. From (2.5),

$$\underline{\underline{S}}_{ik}\,\underline{a}^{(k)} = \sum_{\ell=1}^{4} \gamma_\ell^{(i,k)} \underline{\alpha}_\ell^{(i)} \underline{\beta}_\ell^{(k)T} \underline{a}^{(k)} = \sum_{\ell=1}^{4} \gamma_\ell^{(i,k)} \underline{\alpha}_\ell^{(i)} Z_\ell^{(k)} \tag{3.1}$$

where the scalar $Z_\ell^{(k)}$ is,

$$Z_\ell^{(k)} = \underline{\beta}_\ell^{(k)T} \underline{a}^{(k)} \tag{3.2}$$

Thus (2.2) becomes

$$\underline{\underline{V}}_{ii}\underline{a}^{(i)} = \underline{g}^{(i)} - \sum_{k=i-1}^{i+1} \sum_{\ell=1}^{4} \gamma_\ell^{(i,k)} Z_\ell^{(k)} \underline{\alpha}_\ell^{(i)} \tag{3.3}$$

or

$$\underline{a}^{(i)} = \underline{\tilde{g}}^{(i)} - \sum_{k=i-1}^{i+1} \sum_{\ell=1}^{4} \gamma_\ell^{(i,k)} Z_\ell^{(k)} \underline{\tilde{\alpha}}_\ell^{(i)} \tag{3.4}$$

where

$$\underline{\tilde{g}}^{(i)} = \underline{\underline{V}}_{ii}^{-1}\,\underline{g}^{(i)}, \quad \underline{\tilde{\alpha}}_\ell^{(i)} = \underline{\underline{V}}_{ii}^{-1}\,\underline{\alpha}_\ell^{(i)}$$

Provided the $Z_\ell^{(k)}$ are known, (3.4) gives the solution of (2.2). To find $Z_\ell^{(k)}$, premultiply (3.4) by $\underline{\beta}_m^{(i)T}$ to get

$$Z_m^{(i)} + \sum_{k=i-1}^{i+1} \sum_{\ell=1}^{4} \gamma_\ell^{(i,k)} \underline{\beta}_m^{(i)T} \underline{\tilde{\alpha}}_\ell^{(i)} Z_\ell^{(k)} = \underline{\beta}_m^{(i)T} \underline{\tilde{g}}^{(i)} \tag{3.5}$$

Eq. (3.5) is a set of 4Mx4M equations for the unknown $Z_m^{(i)}$. After assembly and solution of (3.5), the $\underline{a}^{(i)}$ can be found from (3.4).

The algorithm for this "decoupling" approach is thus:

(1) For each element i, i=1, ...,M form $\tilde{\underline{\alpha}}_m^{(i)}$, $\tilde{\underline{g}}^{(i)}$ by an appropriate factorisation technique.

(2) Assemble the auxiliary 4Mx4M system (eq. 3.5) and solve for the $z_m^{(i)}$ by an appropriate factorisation technique.

(3) Compute the desired solution in each element from (3.4).

Further details of the factorisation techniques used can be found in section 4.1.

3.2 Iteration and Decoupling

A straightforward iterative technique can be based on a split of the matrix $\underline{\underline{V}}_{ii}$ into

$$\underline{\underline{V}}_{ii} = \underline{\underline{V}}_{io} + \underline{\underline{\delta V}}_i$$

Writing (3.7)

$$\underline{a}_{n+1}^{(i)} = \underline{a}_n^{(i)} + \underline{\delta a}_{n+1}^{(i)} \qquad n = 0,1,\ldots.$$

$$\underline{a}_o^{(i)} \text{ given}$$

yields the obvious iterative scheme for (2.2)

$$(\underline{\underline{V}}_{io} + \underline{\underline{S}}_{ii})\underline{\delta a}_{n+1}^{(i)} = \underline{g}^{(i)} - \sum_k \underline{\underline{S}}_{ik}\underline{\delta a}_{n+1}^{(k)} - (\underline{\underline{V}}_{ii} + \underline{\underline{S}}_{ii})\underline{a}_n^{(i)} - \sum_k \underline{\underline{S}}_{ik}\underline{a}_n^{(k)}$$

$$= \underline{r}_n^{(i)} - \sum_k \underline{\underline{S}}_{ik}\underline{\delta a}_{n+1}^{(k)} \qquad (3.8)$$

Note that (3.8) is formally similar in structure to (2.2) and thus the technique of section 3.1 can be applied to eq. (3.8) at each stage n of the iteration.

Since the $\underline{\underline{V}}_{ii}$ are asymptotically diagonal (see Delves and Phillips (1980)), an appropriate splitting of $\underline{\underline{V}}_{ii}$ is

$$\left(\underline{\underline{V}}_{io}\right)_{pq} = \begin{cases} \left(\underline{\underline{V}}_{ii}\right)_{pq} & |p-q| \leqslant \mu \\ 0 & \text{otherwise} \end{cases}$$

i.e. $\underline{\underline{V}}_{ii}$ is replaced by a banded matrix $\underline{\underline{V}}_{io}$ of semibandwidth μ.

The iterative/decoupling algorithm then has a preliminary part and an iterative loop:

Preliminary Part

(1) For each element i, form $\tilde{\underline{\alpha}}_m^{(i)} = \underline{\underline{V}}_{i0}^{-1} \underline{\alpha}_m^{(i)}$, by an appropriate factorisation technique (see section 4.2), retaining the factorisation for later use. Note that this factorisation can take place in parallel for each element i since the elements have been decoupled by the low rank modification technique.

(2) Assemble the auxiliary matrix and factors by an appropriate technique retaining the factorisation for later use.

Iterative loop

(1) Form the residuals $\underline{r}_n^{(i)}$ in (3.8).

(2) From the saved factorisation of $\underline{\underline{V}}_{i0}$ compute $\tilde{\underline{r}}_n^{(i)}$

(3) From the saved factorisation of the auxiliary matrix, compute the $Z_m^{(i)}$

(4) Compute the $\underline{\delta a}_{n+1}^{(i)}$

Further details of the factorisation techniques can be found in section 4.2.

4. RESULTS

We have implemented and timed the iteration and decoupling schemes on both the ICL DAP and CDC7600. Throughout, the number of trial functions N in each element has been set at N=64, and in section 4.2, the convergence criterion is

$$\| \underline{\delta a}_{n+1}^{(i)} \| \leqslant 10^{-8} \| \underline{a}_n^{(i)} \|$$

in each element i, i=1,....M

4.1 Decoupling Method

The results for the direct (non-iterative) decoupling method described in section 3.1 are presented in Table 1. On the CDC 7600, the linear equations were solved by the usual Gauss elimination (with column pivoting) to reduce the full matrix to triangular form, while on the DAP, Gauss-Jordan (again with column pivoting) was used to reduce the full matrix to diagonal form. REAL*8 arithmetic was used on the DAP to give a direct comparison with the 7600 60bit results.

As expected, the timings indicate a linear dependence on M on both the CDC 7600 and DAP. The ratio of CDC/DAP times indicates that using this technique (in REAL*8 on the DAP) the DAP is approximately 0.37xCDC7600 in power.

Table 1

Decoupling Results

Number of Elements M	1	2	3	4	5	6	7	8
DAP time msecs (REAL*8)	307	613	920	1226	1533	1840	2146	2453
CDC 7600 Time msecs	112	225	337	450	562	675	788	901
CDC/DAP (REAL*8)	0.36	0.37	0.37	0.37	0.37	0.37	0.37	0.37
DAP time msecs (REAL*4)	123							
CDC/DAP (REAL*4)	0.91							

For comparison, the DAP result for REAL*4 for M=1 is displayed. There is a degradation of about 2.5 in speed when REAL*8 is used in place of REAL*4, so that a speed comparison of DAP REAL*4 to CDC7600 would yield a ratio of about 0.9.

It should be noted that in this decoupling method, there is not much parallelism to be exploited on the DAP, since with N=64 the method essentially reduces to the solution of full sets of 64x64 equations.

4.2 Iteration and Decoupling Method

To successfully use this approach, efficient solvers for banded systems (of arbitrary semibandwidth μ) must be available on both the DAP and CDC 7600. As far as the authors are aware, no such solver exists for a DAP-like architecture. Thus in order to proceed, a slightly different form of $\underline{\underline{V}}_{io}$ (see section 3.2) was used. Here $\underline{\underline{V}}_{io}$ was taken to be block tridiagonal, each block being of size $\mu x \mu$ and on the DAP, parallel cyclic reduction (extended to block tridiagonal systems) was used (see Whiteway (1979), Delves, Samba and Hendry (1983)). Essentially cyclic reduction reduces the matrix after a finite number of operations to block diagonal form which can then be solved directly. Unfortunately it is not possible to save the matrix in situ at each stage of reduction for subsequent use. To bypass this difficulty, the matrices at each stage of the reduction were saved in individual DAP matrices, a technique which proved satisfactory for the model problem considered, but could not be contemplated for more general cases. The cyclic reduction was applied by fitting as many elements as possible into a DAP matrix so that many independent sets of block tridiagonal systems could be solved in parallel (e.g. for μ=8, the equations associated with 8 elements can be solved). No pivoting was used during the cyclic reduction.

On the CDC 7600, $\underline{\underline{V}}_{io}$ can be viewed as a variable banded matrix of semibandwidth varying from μ to 2μ, and Gauss elimination taking

Table 2

Iterative/Decoupling Results (DAP in REAL*8)

Number of Elements M	μ	Iterations	Total Time (msecs)	Preliminary Time (msecs)	Time/Iteration (msecs)	
4	16	5	2704	1003	340	DAP
			531	193	68	CDC
			0.20	0.19	0.20	CDC/DAP
4	14	6	3737	1239	416	DAP
			537	156	64	CDC
			0.14	0.13	0.15	CDC/DAP
5	12	8	4016	1089	366	DAP
			761	161	75	CDC
			0.19	0.15	0.20	CDC/DAP
6	10	13	5065	942	317	DAP
			1248	157	84	CDC
			0.25	0.17	0.26	CDC/DAP
8	8	15 (no convergence)	4661	776	259	DAP
			1724	166	104	CDC
			0.37	0.21	0.40	CDC/DAP

account of the variable bandwidths was used (without pivoting for consistency with the DAP).

The auxiliary matrix systems were handled as in section 4.1 on both the DAP and CDC 7600.

Table 2 shows the results for varying number of elements and bandwidth μ for both the CDC 7600 and DAP. The total times for convergence (or divergence!) are shown together with the times for the preliminary part of the algorithm and a single iteration. Also shown are the ratios of the CDC/DAP times. Throughout REAL*8 has been used on the DAP.

With perhaps the exception of $\mu=8$, all of these ratios are disappointingly low say 2 to 3 times smaller than the ratio in Table 1. Only for $\mu=8$ do the ratios approach that of Table 1, but unfortunately the iteration fails to converge for this value of μ!

5. CONCLUSIONS

The results of section 4.2 indicate that the hoped for improvement in the DAP performance (relative to the CDC 7600) using the iterative technique is not realised. In fact, when measured against the decoupling approach of section 4.1, there is a significant degradation in performance over the CDC 7600.

Many possibilities associated with the iteration algorithm on the DAP could be proposed for this:

(i) The DAP algorithm is inefficiently coded.

(ii) An inappropriate banded solver has been used on the DAP.

(iii) For the model problem chosen, the DAP is being applied to too small a problem.

(iv) The iterative algorithm is based on "too serial" an approach and a more parallel technique for the solution of the GEM equations is needed on the DAP.

We believe that the DAP code has been reasonably well implemented and hence rather discount (i). It should be noted, that even on a serial machine the iterative algorithm is not itself without difficulties, since for the CDC 7600 results, the times for the iteration/decoupling approach are always larger than the corresponding times for the decoupling algorithm. This is due to needing a quite large bandwidth to achieve convergence. This observation suggests that the iteration technique itself is being applied to too small a problem. For $N>>64$, the banded matrices involved would be relatively less full and the iterative technique would then show to better advantage. However it is also, we believe, true that further work on banded solvers for the DAP would be well worthwhile. We cannot discount possibility (iv) either.

ACKNOWLEDGEMENT

The work reported here was supported in part by SERC Grant GR/B/24332 to the Universities of Birmingham and Liverpool.

REFERENCES

Delves, L.M. and Hall, C.A. (1979) An implicit matching procedure for global element calculations, *J. Inst. Math. Applics.*, **23**, pp. 223-234.

Delves, L.M., Samba, A.S. and Hendry, J.A. (1983) Band Matrices on the DAP (this proceedings).

Delves, L.M. and Phillips, C. (1980) A fast implementation of the global element method, *J. Inst. Math. Applics.*, **25**, pp. 177-197.

Hendry, J.A. (1980) Singular problems and the global element method, *Comp. Methods in Appl. Mech. and Engineering*, **21**, pp. 1-15.

Hendry, J.A., Delves, L.M. and Phillips, C. (1978) Numerical experiences with the global element method, in The Mathematics of Finite Elements and its Applications III (J.R. Whiteman, Ed.), Academic Press, pp. 341-348.

Whiteway, J. (1979) A parallel algorithm for solving tridiagonal systems, in *DAP Newsletter*, **3**, pp. 5-8.

THE IMPLEMENTATION OF THE FFT ON THE DAP

S.T. Davies

(DAP Support Unit, Queen Mary College, London)

ABSTRACT

By discussing the basic theory of the FFT it is shown that the FFT algorithm can be implemented in a natural way on the DAP. The 64-point algorithm is considered in detail and the DAP FORTRAN code for calculating this case is presented. Code for calculating the multiplication factors and reordering the data is also included. Ways of modifying this code to perform more general and varied FFTs, including the 2-dimensional case, are discussed. A final short section is included giving some indication of the performance of these routines on the DAP.

1. INTRODUCTION

The intention of this paper is to present, in a very simple way, the ideas and routines involved in implementing particular cases of the Fast Fourier Transform (FFT) on the ICL Distributed Array Processor (DAP). By covering some of the basic theory and the FFT algorithm in detail it is shown that the FFT algorithm can be implemented in an entirely nautral way on the DAP.

2. BASIC THEORY

The Discrete Fourier Transform (DFT) of a set of N complex points $X(j)$ $j = 0,\dots,N-1$ is a set of N complex points $\hat{X}(k)$ $k=0,\dots,N-1$ defined by

$$\hat{X}(k) = \sum_{j=0}^{N-1} X(j).W_N^{kj} \qquad \text{where } W_N = e^{-2\pi i/N} \qquad (1\text{-}1)$$

and the FFT is an algorithm which calculates these sums efficiently for certain forms of N.

The simplest form of FFT is when $N=2^M$ for some M, and is based on the observation that the sequence $X(j)$ can be split into two sequences

$$X_0(j) = X(2j) \qquad j = 0,\dots,N/_2 - 1$$

$$X_1(j) = X(2j + 1) \quad j = 0,\dots,N/_2 - 1$$

and (1-1) can be written

$$\sum_{j=0}^{N-1} X(j).W_n^{kj} = \sum_{j=0}^{N/_2-1} X(2j).W_N^{2kj} + \sum_{j=0}^{N/_2-1} X(2j+1).W_N^{k(2j+1)}$$

$$= \sum_{j=0}^{N/_2-1} X_0(j).W_{N/_2}^{kj} + W_N^k \sum_{j=0}^{N/_2-1} X_1(j).W_{N/_2}^{kj}$$

for $k=0,\ldots,N/_2 - 1$ the sums on the r.h.s. are again DFTs so we have

$$\hat{X}(k) = \hat{X}_0(k) + W_N^k.\hat{X}_1(k) \qquad k = 0,\ldots,N/_2 - 1.$$

Looking at the range $k = N/_2,\ldots,N-1$ we find that

$$\hat{X}(N/_2 + k) = \hat{X}_0(k) - W_N^k.\hat{X}_1(k) \qquad k = 0,\ldots,N/_2 - 1.$$

Thus the original DFT of N points has been reduced to the calculation of two $N/_2$ - point DFTs but with the introduction of the factor W_N^k. This process can now be repeated on each of the sequences $X_0(j), X_1(j)$ to reduce the problem to the calculation of four $N/_4$ - point DFTs again with the introduction of certain multiplication factors. We can continue in this way (M steps) until we have to calculate N 1-point transforms and the DFT of a single point is the point itself. The FFT algorithm works by going through this process in reverse and forms the DFT as a series of sums in the form $T_1 + W.T_2$ where T_1 and T_2 are previously calculated transforms and W is a multiplying factor.

3. A LOOK AT THE 8-POINT FFT

The steps described above may be written formally for the 8-point case as follows:

Step 1 $\quad \hat{X}(k) = \hat{X}_0^{(1)}(k) + W_8^k.\hat{X}_1^{(1)}(k) \qquad k = 0,1,2,3$

$\hat{X}(k+4) = \hat{X}_0^{(1)}(k) - W_8^k.\hat{X}_1^{(1)}(k) \qquad$ "

Step 2

$$\hat{X}_0^{(1)}(k) = \hat{X}_0^{(2)}(k) + W_4^k.\hat{X}_1^{(2)}(k) \qquad k = 0,1$$

$$\hat{X}_0^{(1)}(k+2) = \hat{X}_0^{(2)}(k) - W_4^k.\hat{X}_1^{(2)}(k) \qquad "$$

$$\hat{X}_1^{(1)}(k) = \hat{X}_2^{(2)}(k) + W_4^k.\hat{X}_3^{(2)}(k) \qquad k = 0,1$$

$$\hat{X}_1^{(1)}(k+2) = \hat{X}_2^{(2)}(k) - W_4^k.\hat{X}_3^{(2)}(k) \qquad "$$

Step 3

$$\hat{X}_0^{(2)}(0) = \hat{X}_0^{(3)} + W_2^0.\hat{X}_1^{(3)}$$

$$\hat{X}_0^{(2)}(1) = \hat{X}_0^{(3)} - W_2^0.\hat{X}_1^{(3)}$$

$$\hat{X}_1^{(2)}(0) = \hat{X}_2^{(3)} + W_2^0.\hat{X}_3^{(3)}$$

$$\hat{X}_1^{(2)}(1) = \hat{X}_2^{(3)} - W_2^0.\hat{X}_3^{(3)}$$

$$\hat{X}_2^{(2)}(0) = \hat{X}_4^{(3)} + W_2^0.\hat{X}_5^{(3)}$$

$$\hat{X}_2^{(2)}(1) = \hat{X}_4^{(3)} - W_2^0.\hat{X}_5^{(3)}$$

$$\hat{X}_3^{(2)}(0) = \hat{X}_6^{(3)} + W_2^0\,\hat{X}_7^{(3)}$$

$$\hat{X}_3^{(2)}(1) = \hat{X}_6^{(3)} - W_2^0\,\hat{X}_7^{(3)}$$

where $\hat{X}_j^{(i)}$ denotes the j'th transform at step i.

At step 3 we require the single point transforms $\hat{X}_0^{(3)}, \hat{X}_1^{(3)}, \ldots, \hat{X}_7^{(3)}$ and working through the way the sums are split up we see that

$$\hat{X}_0^{(3)} = X(0) \quad \hat{X}_1^{(3)} = X(4) \quad \hat{X}_2^{(3)} = X(2) \quad \hat{X}_3^{(3)} = X(6)$$

$$\hat{X}_4^{(3)} = X(1) \quad \hat{X}_5^{(3)} = X(5) \quad \hat{X}_6^{(3)} = X(3) \quad \hat{X}_7^{(3)} = X(7).$$

We can now work backwards through these steps as displayed in the diagram (signal flow graph). (Fig. 1)

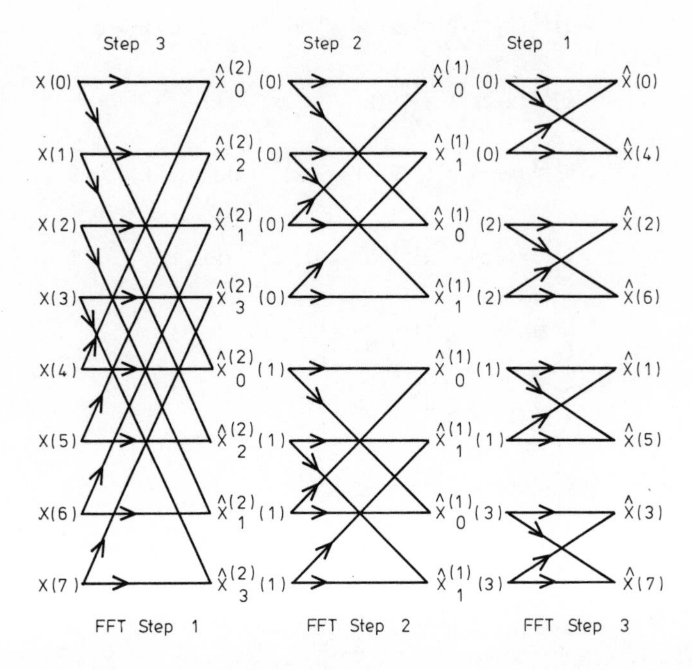

Fig. 1

The lines connecting the nodes are intended to show which data items are required at each node at each step. We can see from the diagram that the operations involved in the FFT are essentially shifts, multiplies and adds which can be performed largely in parallel.

If we suppose the data X(j) j=0,...,7 are contained in a vector X and we have vectors W_i i = 1,2,3 containing the relevant multiplication factors, the parallel FFT algorithm (using complex operations) looks like:

```
1     k = 4
2     do 1 i = 1,3
3     set MASK = alternating pattern of k false, k true, k false......
4     set XT = shift left of X by k
5     where MASK is true set XT = -X
6     where MASK is true set X = shift right of X by k
7     set XT = XT*W_i
8     set X = X + XT
9     k = k/2
10    1 continue
```

After the three steps the vector X contains the component X(j) j = 0,...,7 in the order shown in the diagram. This algorithm is the basis of most FFT routines on the DAP. The shifts in line 4 and 6 correspond to the diagonal data connections in the diagram and line 5 takes care of the alternating signs occurring in the system of equations above. To illustrate this consider step 1 of the loop which runs as follows:

X =	X(0)	X(1)	X(2)	X(3)	X(4)	X(5)	X(6)	X(7)	
k = 4									(line 1)
MASK =	F	F	F	F	T	T	T	T	(line 3)
XT =	X(4)	X(5)	X(6)	X(7)	0	0	0	0	(line 4)
XT =	X(4)	X(5)	X(6)	X(7)	-X(4)	-X(5)	-X(6)	-X(7)	(line 5)
X =	X(0)	X(1)	X(2)	X(3)	X(0)	X(1)	X(2)	X(3)	(line 6)
XT =	W_1 *XT								(line 7)
X =	X + XT								(line 8)

The remaining two steps of the loop work in the same way on the new data but using shift distances of 2 and 1, and the vectors of multiplying factors W_2 and W_3 respectively.

Looking at the vectors of multiplication factors W_i we see that

$$W_1 = (W_2^0,W_2^0,W_2^0,W_2^0,W_2^0,W_2^0,W_2^0,W_2^0,) = (1,1,1,1,1,1,1,1,)$$

$$W_2 = (W_4^0,W_4^0,W_4^0,W_4^0,W_4^1,W_4^1,W_4^1,W_4^1,) = (1,1,1,1,-i,-i,-i,-i)$$

$$W_3 = (W_8^0,W_8^0,W_8^2,W_8^2,W_8^1,W_8^1,W_8^3,W_8^3,) = (1,1,-i,-i,W_8^1,W_8^1,-iW_8^1,-iW_8^1)$$

Thus the multiplications in line 7 of the algorithms are very simple. Step 1 requires no multiplication at all. In Step 2 we have to multiply some components by -i and since -i(a+ib)=b-ia we have again no multiplication but only masked assignment and negation. In Step 3, in addition to multiplication by -i, we have to multiply some components by $W_8^1 = e^{-i\pi/4} = \frac{1}{\sqrt{2}}(1-i)$ and since

$$\frac{1}{\sqrt{2}}(1-i)(a+ib) = \frac{1}{\sqrt{2}}((a+b)+i(-a+b))$$

we need only do two additions and a multiplication by $\frac{1}{\sqrt{2}}$. The operations just described are performed on various alternating groups of components and the relevant masks required for these operations can be obtained by simply keeping copies of the MASKs generated at each step. So by doing each loop of the algorithm explicitly we have, in principle, an efficient way of implementing the 8-point FFT.

As remarked above the components of the transform appear in the order shown in the diagram. This is the so called 'bit reversed' order, so that, for example, location 1 (=001) contains component 4 (=100) of the transform. It may thus be necessary to reshuffle the result - more will be said of this later.

4. THE 64-POINT FFT

Clearly the above procedure could have been carried through for a 64-point transform, this time requiring six vectors W_i $i=1,\ldots,6$ of multiplication factors, but now in steps 4 to 6, the multiplication corresponding to line 7 of the algorithm are full complex multiplications.

We can however, improve on this procedure by noting 64 = 8 x 8 so that in (1-1) we can write

$$k = 8k_1 + k_0 \qquad k_1, k_0 = 0, \ldots, 7$$

$$j = 8j_1 + j_0 \qquad j_1, j_0 = 0, \ldots, 7$$

and writing

$$X(j) = X(j_1, j_0)$$

$$\hat{X}(k) = \hat{X}(k_1, k_0)$$

the sum (1-1) can be arranged as follows

$$\hat{X}(k) = \hat{X}(k_1, k_0) = \sum_{j=0}^{63} X(j).W_{64}^{jk}$$

$$= \sum_{j_0=0}^{7} \sum_{j_1=0}^{7} X(j_1, j_0).W_{64}^{(8j_1 + j_0)(8k_1 + k_0)}$$

$$= \sum_{j_0=0}^{7} \sum_{j_1=0}^{7} X(j_1, j_0).W_{64}^{(64j_1k_1 + 8j_1k_0 + 8j_0k_1 + j_0k_0)}$$

$$= \sum_{j_0=0}^{7} \sum_{j_1=0}^{7} X(j_1, j_0).W_8^{j_1k_0}.W_{64}^{j_0k_0}.W_8^{j_0k_1}$$

since $W_{64}^{64k_1j_1} = 1$, $W_{64}^{8j_1k_0} = W_8^{j_1k_0}$, $W_{64}^{8j_0k_1} = W_8^{j_0k_1}$.

The final double summation can be broken down into the following three steps

$$\text{(a)}\quad X'(j_0,k_0) = \sum_{j_1=0}^{7} X(j_1,j_0).W_8^{j_1k_0}$$

$$\text{(b)}\quad X''(j_0,k_0) = X'(j_0,k_0).W_{64}^{j_0k_0}$$

$$\text{(c)}\quad \hat{X}(k_1,k_0) = \sum_{j_0=0}^{7} X''(j_0,k_0).W_8^{j_0k_1}.$$

Note that for each fixed j_0 the sum in (a) has the form of an 8-point DFT - each DFT being formed from points of X which are separated by 8 positions. So for $j_0 = 0$ points X(0), X(8), X(16),...,X(56) are used and for $j_0 = 1$ points X(1), X(9),...,X(57) are used, and so on. We can now use the 8-point FFT algorithm above to evaluate the eight DFTs in (a) in parallel, but because the points are separated by 8 positions we require shifts of 32,16 and 8 rather than 4,2 and 1. In step (b) we simply have to multiply the $X(j_0,k_0)$ by the factors $W_{64}^{j_0k_0}$ $j_0,k_0 = 0,\ldots,7$, but having done the DFTs in (a) the $X(j_0,k_0)$ will appear in bit-reversed order in the k_0 index and so the $W_{64}^{j_0k_0}$ must also appear with the k_0 index bit-reversed. Again, for fixed k_0, the sums in (c) have the form of 8-point DFT - each transform, this time, being formed from groups of eight consecutive points, and we can use the 8-point FFT algorithm to perform these transforms in parallel. Thus the 64-point transform can be broken down into two interleaved sets of eight 8-point transforms with a multiplication by suitable factors between the two sets of transforms. (Those familiar with FFTs will recognise the above as a particular case of the "twiddle factor" representation of FFTs, the $W_{64}^{j_0k_0}$ being the twiddle factors.)

If we now suppose the vector X consists of vectors of real and imaginary parts XR, XI and we have vectors WR, WI containing the real and imaginary parts of the factors $W_{64}^{j_0k_0}$ in the correct order we can write the DAP FORTRAN code to perform the 64-point transform efficiently using the algorithm described for the eight-point transform as follows:

```
      SUBROUTINE VFFT(XR,XI)
C     REAL XR(),XI(),XRS(),XIS(),ZS()
      REAL WR(),WI()
      COMMON / FACTORS/WR,WI
      LOGICAL M1(),M2(),M3(),KLOG()
      EQUIVALENCE (KLOG,K)
```

```
C
      RROOT2=0.7071068
C
C Steps 1&4 : multipliers are all 1
C
      KLOG =.FALSE.
      K=32
1     XRS = SHLP(XR,K)
      XIS = SHLP(XI,K)
      MI = ALT(K)
      XRS(M1) = -XR
      XIS(M1)= -XI
      XR(M1) = SHRP(XR,K)
      XI(M1) = SHRP(XI,K)
      XR = XR + XRS
      XI = XI + XIS
      KLOG = SHRP(KLOG)
C
C Steps 2&5 : multipliers are 1 and -i
C
      XRS = SHLP(XR,K)
      XIS = SHLP(XI,K)
      M2 = ALT(K)
      XRS(M2) = -XR
      XIS(M2) = -XI
      XR(M2) = SHRP(XR,K)
      XI(M2) = SHRP(XI,K)
      XI(M2) = SHRP(XI,K)
C
C Multiply by -i in the second half.
C
      ZS = XIS
      XIS(M1) = -XRS
      XRS(M1) = ZS
C
C Then add.
```

```
C
      XR = XR + XRS
      XI = XI + XIS
      KLOG = SHRP(KLOG)
C
C Steps 3&6 : multipliers are 1, -i and (1-i)/sqrt(2)
C
      XRS = SHLP(XR,K)
      XIS = SHLP(XI,K)
      M3 = ALT(K)
      XRS(M3) = -XR
      XIS(M3) = -XI
      XR(M3) = SHRP (XR,K)
      XI(M3) = SHRP(XI,K)
C
C In second half multiply by (1-i)/sqrt(2)
      ZS(M1) = -XRS + XIS
      XRS(M1) = (XRS+XIS)*RROOT2
      XIS(M1) = ZS*RROOT2
C
C In alternate quarters multiply by -i.
C
      ZS = XIS
      XIS(M2) = -XRS
      XRS(M2) = ZS
C
C Then add.
C
      XR = XR + XRS
      XI = XI + XIS
      KLOG = SHRP(KLOG)
C
C Prepare for step 4 by multiplying by suitable factors.
```

```
C
      IF(K .LT. 4)RETURN
      ZS = XR*WI
      XR = XR*WR - XI*WI
      XI = XI*WR + ZS
      GOTO 1
C
      END
```

4.1 Calculation of the W factors

We need to compute the vectors WR and WI containing the real and imaginary parts of the $W_{64}^{j_o k_o}$ $j_o, k_o = 0,\dots,7$ with the k_o index bit-reversed and referring to elements separated by 8 positions. We can write the j_o and the k_o indices in terms of two vectors JO,KO whose product will give the required combinations of j_o and k_o in the correct order. The vectors are

```
JO: 0 1 2...7 0 1 2...7 0 1...7 0 1...7 0 1...7 0 1...7 0 1...7 0 1...7
KO: 0 0 0...0 4 4 4...4 2 2...2 6 6...6 1 1...1 5 5...5 3 3...3 7 7...7
```

The JO vector simply repeats groups of eight consecutive numbers 0 to 7. The KO vector repeats groups of the eight numbers 0 to 7 in bit-reversed order and separated by eight positions.

The vectors WR and WI are then given by

```
        THETA = 2 * π *JO*KO/64.0        (THETA a real vector)
        WR = COS(THETA)
        WI = -SIN(THETA)
```

If we consider the bit patterns of the numbers in the vectors JO,KO we see that each number requires at most three bits and that the vectors can be constructed from logical vectors as follows

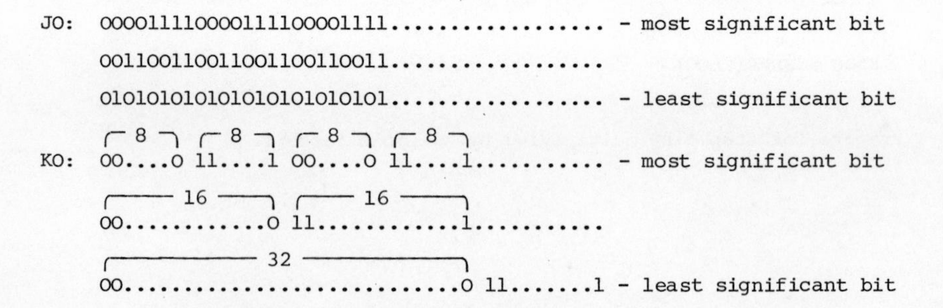

We thus take advantage of DAP FORTRAN logical variables to write the DAP FORTRAN code for computing the vectors WR and WI.

```
SUBROUTINE FACS
REAL WR(),WI()
INTEGER JO(),KO()
LOGICAL LJO(,),LKO(,)
COMMON /FACTORS/WR,WI
EQUIVALENCE (JO,LJO),(KO,LKO)
JO=0
KO=0
LJO(,64)=ALT(1)
LJO(,63)=ALT(2)
LJO(,62)=ALT(4)
LKO(,64)=ALT(32)
LKO(,63)=ALT(16)
LKO(,62)=ALT(8)
WR=2.0*3.14159*JO*KO/64.0
WI=-SIN(WR)
WR=COS(WR)
RETURN
END
```

4.2 Shuffling the data

It was noted above that the results of the 8-point FFT appear in bit-reversed order. The same is true of the 64-point transform with the bit-reversal now being on six bits, so for example location 1 (=000001) will contain component 32 (=100000) of the transform. The data can be shuffled efficiently on the DAP using shifts and masked assignments, to give the results in the same order as the input data. The code to do this for the 64-point case is as follows.

```
      SUBROUTINE VBREV(XR,XI)
C
      REAL XR(),XI(),XRS(),XIS()
      LOGICAL M(),KLOG(),NLOG()
      EQUIVALENCE (KLOG,K),(NLOG,N)
      KLOG=.FALSE
      K=32
      N=1
```

```
  1 M=.NOT.ALT(K).AND.ALT(N)
    IPL=K-N
    XRS=SHRC(XR,IPL)
    XIS=SHRC(XI,IPL)
    XR(M)=SHLC(XR,IPL)
    XI(M)=SHLC(XI,IPL)
    M=ALT(K).AND..NOT.ALT(N)
    XR(M)=XRS
    XI(M)=XIS
C
    KLOG=SHRP(KLOG)
    NLOG=SHLP(NLOG)
    IF(K.NE.4)GO TO 1
    RETURN
    END
```

This code is essentially an implementation of a particular case of the data movement algorithms described in Flanders (1981). The reader may like to verify that the code actually performs the required shuffling.

5. FURTHER DEVELOPMENTS

The routines above were written using DAP FORTRAN vectors in order to calculate a 64-point FFT. This was done largely for simplicity as the vector version obviously does not fully exploit the DAP. However, with virtually identical code, we can actually calculate 64 64-point transforms simultaneously by using MATRIX mode and performing independent transforms either along the rows or down the columns of the matrix. In fact the only changes that have to be made are to change the vectors to matrices and, in the case of row transforms for example, replace SHLP by SHWP, SHRP by SHEP and ALT by ALTC (column transforms would require SHNP, SHSP, and ALTR). The matrices of W factors are either MATC or MATR of the vectors WR and WI (or could have been set up using bit planes generated by ALTC or ALTR), and similar remarks apply to the data shuffling routine.

By performing first row and then column FFTs independently on the input data matrices, we obtain a 64 x 64 2-dimensional FFT, simply using the routines described above.

A further possibility is to note that 4096 = 64 x 64 and that by combining the row and column transforms with multiplication by suitable factors ($W_{4096}^{j_0 k_0}$ $j_0, k_0 = 0, \ldots, 63$ with the k_0 index bit-reversed) between the two transforms we can obtain a 4096-point transform (in long vector order say) in the same way that the 64-point transform is obtained by combining 8-point transforms. The factors in this case can again be set up by making use of logical matrices to set up the various bit-planes of the j_0 and k_0 indices.

Various other power of 2 transforms are possible and a number of these may be performed simultaneously e.g. four 32 x 32 2-dimensional transforms or two 2048-point transforms and so on. Essentially the same code as described above may be used for these cases but fewer steps of the algorithm are executed.

Transforms of numbers of points other than a power of 2 are also possible, and provided we are able to pack the data fairly economically into the DAP and in a way which does not require too much data movement, a wide variety of FFT calculations is possible. As an example, a set of routines have been written for a particular application to calculate four 24 x 24 point 2-dimensional DFTs simultaneously. The data are stored in the obvious 2 x 2 partition of the DAP and despite about 40% of the processors being idle the routines performed satisfactorily.

One possibility not discussed in this paper is that of performing 4096 N-point DFTs simultaneously with the data, in this case, being stored "vertically" in each processing element of the DAP. This technique exploits the parallelism of the DAP in a different way from the technique described above and the algorithm in this case is exactly the same as the serial algorithm.

6. PERFORMANCE

In the 64 x 64 2D case, setting up the multiplication factors takes about 4.2 ms and the actual FFT calculation including the data shuffling, takes about 19 ms. In the 4096-point transform, setting up the factors takes about 6 ms and the FFT, including data shuffling, takes about 21.3ms.

ACKNOWLEDGEMENTS

Many of the ideas in this paper are the result of the work of Professor D. Parkinson (ICL and QMC) and P.M. Flanders (ICL).

REFERENCES

Flanders, P.M. (1981) Musical bits - a generalised method for a class of data movements on the DAP. ICL report CM 70.

QU FACTORISATION AND SINGULAR VALUE DECOMPOSITION ON THE DAP

J.J. Modi

(Computing Laboratory, University of Oxford)

and

G.S.J. Bowgen

(DAP Support Unit, Queen Mary College, London)

1. INTRODUCTION

The Singular Value Decomposition (SVD) introduced by Hestenes (1958), decomposes a real m x n matrix into $U\Sigma V^t$ where U and V are orthogonal matrices of order m x n and n x n respectively and Σ is an n x n diagonal matrix. The method suggested was the application of plane rotations on the matrix A. However, this was superseded by a suggestion of Golub and Kahan (1965), in which Householder transformations reduce A to bi-diagonal form and the QR algorithm is then used.

In this paper we start by considering the parallel implementation of QU factorization and then two parallel methods for accomplishing the SVD are presented.

The standard Givens ordering can be expressed as a set of rotations which being independent can be computed simultaneously (Gentleman (1975), Sameh and Kuck (1978)). We consider the implementation of their ordering. For $m,n \leq 64$ the whole of A fits into DAP store. For larger matrices, special treatment is needed.

An algorithm has been developed which systematically minimises the organisational overheads. The method uses the standard Givens ordering and can be applied to a matrix of any size.

A different Givens ordering, scheme A, suggested by Clarke and Modi (1982) requires fewer parallel steps, but for their ordering the organisational overheads are found to be significantly larger on the DAP.

The first SVD algorithm (referred to as SVD1) is an extension of the parallel Jacobi method. (Modi and Parkinson (1982)). The idea is similar to that of Hestenes in that A is multiplied from the right by orthogonal transformations to diagonalise A^TA. A sequence of matrices $\{P^{(1)} \ldots . P^{(\ell)}\}$, when applied to the right of A, produces a matrix $B=U\Sigma$ whose columns are orthogonal.

The execution time of an algorithm is strongly dependent on the orderings in which the transformations $\{P^{(i)}\}$ are applied. A study to minimise organisational time, for the parallel Jacobi method, on a topological network such as the DAP has been carried out by Modi and Pryce (1982). We have used their orderings.

Luk (1980) has implemented a similar SVD algorithm on the Illiac IV. In this paper a comparison is made with the published results for the Illiac IV and the standard Golub-Reinsch algorithm (using the NAG routine when executing on the ICL 2980 host computer). It turns out that SVD1 is faster than the others, but operates iteratively on the entire plane of the matrix A.

In SVD2, A is transformed into QU, where $U=\begin{bmatrix}\tilde{U}\\ O\end{bmatrix}$ (Q is m x m, U is m x n, and $\tilde{U}$ is n x n upper triangular), and the SVD of $\tilde{U}$ is then found. For the latter step SVD1 can be used efficiently as $\tilde{U}$ is fairly small. SVD2 turns out to be much faster than SVD1.

A modification of the standard error analysis shows that both methods are numerically stable, but the convergence properties of the special cyclic ordering do not appear to carry-over to the General Cyclic Ordering. Convergence of the latter, however, can be guaranteed by angle restriction (Henrici (1958)), Brodlie and Powell (1975)). We here show that for the parallel method based on General Cyclic Ordering, the convergence is also guaranteed by a threshold strategy.

The techniques discussed above have been incorporated in the DAP-FORTRAN Subroutine Library at QMC, and are available for general use.

2. QU FACTORIZATION OF M X N MATRICES ON THE DAP

2.1 Definition of the QU Factorization

Given a real matrix A of dimension m x n (with $m \geq n$) it can be decomposed as

$$A = Q^T U \tag{2.1}$$

where Q^T is an m x m orthogonal matrix and $U = \begin{bmatrix}\tilde{U}\\ O\end{bmatrix}$ with $\tilde{U}$ an n x n upper triangular matrix. The standard Givens' process for determining U involves calculating QA by choosing Q to be a product of plane rotations each of which annihilates an element of A below the main diagonal without destroying the previously introduced zeros. Thus

$$Q = Q_{n-1} \cdots\cdots\cdots Q_2\, Q_1 \tag{2.2}$$

where

$$Q_j = P_j^{(j)} \cdots\cdots P_{m-2}^{(j)}\, P_{m-1}^{(j)} \qquad 1 \leq j \leq n-1 \tag{2.3}$$

and

$$P_i^{(j)} = \begin{bmatrix} 1 & & & & & & O \\ & 1 & & & & & \\ & & \ddots & & & & \\ & & & c_i & s_i & & \\ & & & -s_i & c_i & & \\ & & & & & \ddots & \\ O & & & & & & 1 \end{bmatrix} \begin{matrix} \\ \\ \\ \longleftarrow \text{row } i \\ \longleftarrow \text{row } i+1 \\ \\ \\ \end{matrix} \tag{2.4}$$

This means that $P_i^{(j)}$ annihilates the element at position $(i + 1, j)$.

To perform a single plane rotation, say that given by (2.4), the work involved is as follows. The only rows affected are i and i + 1 and we premultiply these by the matrix

$$P = \begin{bmatrix} c & s \\ -s & c \end{bmatrix} \tag{2.5}$$

i.e. we require

$$\begin{bmatrix} c & s \\ -s & c \end{bmatrix} \begin{bmatrix} 0 & 0 \ldots 0 & u_j & u_{j+1} \ldots u_n \\ 0 & 0 \ldots 0 & v_j & v_{j+1} \ldots v_n \end{bmatrix}$$

where c and s are determined in such a way as to annihilate v_j and satisfy the condition

$$c^2 + s^2 = 1 \tag{2.6}$$

if we set

$$w = (u_j^2 + v_j^2)^{\frac{1}{2}} \tag{2.7}$$

then we find

$$\left.\begin{aligned} \bar{u}_k &= cu_k + sv_k \\ \bar{v}_k &= -su_k + cv_k \end{aligned}\right\} \quad j < k < n \tag{2.8}$$

where

$$c = \frac{u_j}{w} \qquad s = \frac{v_j}{w} \tag{2.9}$$

and $\bar{\underset{\sim}{u}}_k^T$, $\bar{\underset{\sim}{v}}_k^T$ are the updated rows.

From (2.3) we see that Q_j annihilates all elements in column j below the diagonal, sequentially. Sameh and Kuck (1978) have noted that the rotations may be grouped into independent sets and this is more amenable to a parallel computer. The order of annihilation they propose is illustrated in Fig. 2.1(a) for a 24 x 8 matrix. As can be seen the number of eliminations is 30 as opposed to 156 for the serial version. Q can now be written as

$$Q = Q_N \, Q_{N-1} \cdots Q_1 \tag{2.10}$$

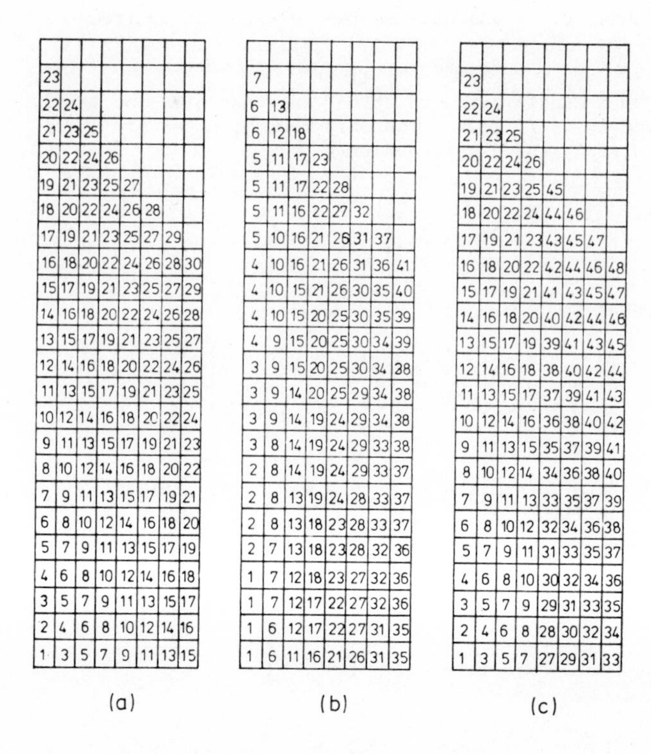

Fig. 2.1 Annihilation orderings

with

$$Q_j = P^{(j)}_{m(j)} \ldots . P^{(j)}_{1} \qquad 1 \leqslant j \leqslant N \tag{2.11}$$

where N is the number of parallel annihilations and m(j) is the number of elements simultaneously eliminated at step j.

2.2 Parallel QU Algorithm

2.2.1 Order of Annihilation on the DAP

We can now consider how to perform QU factorizations on the DAP and for the moment we assume $n \leqslant$ DAPSIZE. If m and n are both less than or equal to DAPSIZE then A can be held completely in one DAP matrix and the annihilation order of Sameh and Kuck (1978) can be used directly. If m > DAPSIZE then we must partition A and use 2 or more DAP matrices to hold it. The obvious partition is simply to slice A into sub-matrices each having DAPSIZE rows. If we assume DAPSIZE is 8 for the rest of this discussion then the slicing is illustrated in Fig. 2.1 for a 24 x 8 matrix A.

To annihilate any particular element involves operations on 2 rows of the matrix and so the maximum number of annihilations that can be performed simultaneously on the DAP is DAPSIZE/2. If we consider Sameh and Kuck's annihilation scheme (Fig. 2.1(a)) then clearly annihilations 9 to 26 all eliminate more than 4 elements simultaneously. In such cases we could carry out the annihilations in 2 stages and so although the number of parallel eliminations required is theoretically 30, on the DAP we must perform 48 such operations to achieve the result. Since the DAP can eliminate at most DAPSIZE/2 elements simultaneously the minimum number of steps needed for this problem is 156/4 = 39 (since there are 156 elements to eliminate). Can we find an annihilation scheme which comes closer to this optimum?

In Fig. 2.1(b) we show a scheme suggested by Clarke and Modi (1982) which requires only 41 steps on the 8x8 DAP. However it is difficult to generate the annihilation pattern for this scheme and more seriously there are complicated data movement overheads involved.

Since we can only operate on DAPSIZE rows of the matrix at any one time some movement of data is inevitable. Suppose we have a work matrix W into which we copy those rows to be operated upon, then the data movement will be simplest if, for any particular slice of the original matrix A, the rows of that slice map directly onto the rows of W. If we consider the sequence of patterns of elements to be annihilated and elements used to eliminate them we find that Clarke and Modi's scheme does not satisfy this condition. In Fig. 2.2 elements to be eliminated are numbered according to the order of elimination and their positions within a DAP matrix are shown. Numbers marked with an asterisk are the elements used to perform the annihilations and it can be seen that at steps 7, 11, 12 and 13 the positions overlap. It is not obvious how these clashes may be resolved in the general case.

1*	2	3*	4	5*	6*	7* 7*	8	9*	10			13*
1*	2	3*	4	5*	6*	7 7*	8	9*	10	11*	12*	13 13*
1*	2	3*	4	5*	6	7*	8	9*	10	11*	12* 12*	13 13
1*	2	3*	4	5*	6	7	8*	9	10*	11*	12* 12	13
1	2*	3	4*	5	6*	7	8*	9	10*	11	12	
1	2*	3	4*	5	6*	7	8*	9	10*	11	12	
1	2*	3	4*	5	6		8*	9	10*	11 11*	12	13*
1	2*	3	4*	5	6		8	9*	10	11		13*

Fig. 2.2 Annihilation clashes of scheme 2.1(b)

Scheme 2.1(a) does not suffer from this drawback and we could use it to factorize A without any row shuffling. However it turns out that an equivalent scheme (Fig. 2.1(c)) is easier to implement. Scheme 2.1(c) is identical to scheme 2.1(a) but the annihilation order has been tailored to suit the DAP.

2.2.2 Algorithm Details

Case $n \leq 64$

We can now consider the implementation details of annihilation scheme 2.1(c) for the DAP when the matrix A is of order $m \times n$ and $n \leq 64$. A is partitioned using the obvious slicing into $[(m-1)/64] + 1$ DAP matrices (so that if m is not an exact multiple of 64 the last matrix in the set will be only partially filled). As described earlier the factorization takes place in 2 stages, eliminating elements in columns 1 to 32 first and then columns 33 to 64 (of course if $n \leq 32$ only 1 stage is needed). For each stage the DAP implementation falls into 3 sections - the start-up phase, the main elimination and the stopping phase. For the 8 x 8 example shown in 2.1(c) these phases are:

start-up phase	steps 1-7	and 27-33
main elimination	steps 8-15	and 34-41
stopping phase	steps 16-26	and 42-48

when $m \leq 64$, the standard annihilation scheme (Fig. 2.1(a)) can be used. The organisation is almost identical to that described below for the start-up phase and stopping phase with the main elimination being omitted.

We will now consider each phase in turn. In order to annihilate 32 elements simultaneously we need the 64 rows concerned to be held in a DAP matrix. To do this we use a work matrix (W) into which the last 64 rows of A are copied at the beginning of the routine. New rows are then brought into W one at a time as required. A logical matrix mask (NEXTROW) is used to index the next row needed and each row of A is copied directly from its position in a slice of A to the corresponding position in W (so no data shifting is needed):

```
NEXTROW = SHNC(NEXTROW)
A(NEXTROW, SLICE + 1) = W
W(NEXTROW) = A(,,SLICE)
```

Throughout the factorization we use a pair of logical masks, MASKV and MASKU, to indicate respectively the elements being annihilated and those in the preceding row used to define the angle of rotation. However during the start-up phase less than 32 elements are being annihilated at each step (see Fig. 2.1(c)) and another logical mask (ACTIVE-ROWS) is used to define which rows have entered the calculation at any particular moment. Once all the rows are "active" the first phase is complete and the main elimination begins.

The main elimination is essentially straight forward. At each step a new row is copied from A into the work matrix and then one parallel annihilation takes place. The arithmetic of equations (2.7)-(2.9) is carried out in the subroutine ANNIHILATE. Equation (2.7) is evaluated in 3 parallel operations and equation (2.9) can be performed with a single parallel division (masking out possible zero divides if $w=0$). Values for c and s are extracted using a DAP indexing technique which picks one component from each row (as indicated by a logical matrix) and are then spread along the rows ready for use in evaluating equation (2.8). Finally (2.8) requires 2 multiplies and 1 addition (the unitary minus needed in the second equation taking a negligible amount of time) ANNIHILATE takes about 1.5 ms to execute.

The stopping phase begins as soon as we need to use row 64 of A (i.e. the last row of the first matrix slice of A). We must ensure that the program stops copying rows of A into W once all the usable ones have been taken. This is done by comparing the NEXTROW mask with another logical matrix indicating the FINAL ROW to be used. The next problem is that we must not attempt to annihilate elements in the final upper triangle, and we can use a logical mask (NON-CAND) to prevent this. The NON-CAND mask is simply ANDed with the set of possible candidates for annihilation until this results in a completely .FALSE. matrix.

Modifications required for the case $n > 64$

If the matrix A is more than 64 columns wide it is necessary to use more than one "column" of DAP matrices to store it. Again, a sliced mapping is the obvious method to use and a 64K x 64L matrix can be stored in a DAP matrix array declared as A(,,K,L). The program changes needed to implement the general case are fairly few. Essentially we have to introduce some DO loops to run over the "column" subscript of the DAP matrices, and the work matrix W becomes a work matrix array of size L.

2.2.3 Timings

In this section we produce some estimates for the time required to factorize matrices of different sizes and compare the actual times taken on the DAP with those on other machines. The time taken to perform one call to the annihilation subroutine (which eliminates 32 elements simultaneously) is 1.5 ms. Therefore the time to annihilate a whole matrix will be 128 x 1.5 ms and with organisational overheads this comes to about 0.22 seconds.

The cost of annihilating a 64K x 64L matrix can be estimated by considering the work done as follows. If we define the time taken to annihilate a whole DAP matrix to be 2 units, then in the case $L = 1$ the total time will be approximately 2K units. When $L = 2$, the time to annihilate the first column of DAP matrices will be 2K + K (since the rotation angles are only calculated once and take roughly half the time of a matrix annihilation). Annihilation of the second column then takes 2(K-1) units (since no work is done on the top matrix in the column). Extending these ideas we can show by induction that the time required for a 64K x 64L matrix will be approximately

$$t = \left[\frac{L^2 + 3L}{2}\right] K - \sum_{i=1}^{L-1} \frac{i^2 + 3i}{2} \tag{2.12}$$

units. From above we see that one unit will be about 0.11 second. Table 2.1 summarises the timings for various problem sizes when run on the DAP, and ICL 2980 (32 bit precision). It also gives the estimated DAP time (using (2.12)) and this can be seen to be a slight over-estimate as K and L get large. The 2980 times are based on the

NAG routine F01QAF. Using 64 bit precision on the DAP would increase the times by a factor of about 3.

The DAP time for the 64 x 64 case is less than the estimate because, as mentioned earlier, this special case is treated slightly differently using the standard annihilation scheme (Fig. 2.1(a)).

Table 2.1

QU Factorization times for various sizes of problem. All times are in seconds

Problem size (64 x 64 blocks)	2980	DAP	DAP estimate (equation 2.12)
1 x 1	1.3	0.19	0.22
2 x 2	9.5	0.88	0.88
3 x 1	4.7	0.66	0.66
3 x 2	16.1	1.5	1.43
3 x 3	31	2.3	2.2
10 x 1	16.5	2.2	2.2
10 x 2	62	5.25	5.28
10 x 3	134	9.2	9.13

2.3 Square Root Free Givens' Transformations

The use of Givens' rotations for QU factorization has been encouraged on serial processors by the development of the so called "square root free" Givens' transformation. Originally introduced by Gentleman (1973), this technique reduces the work required by a considerable amount and removes the need for the calculation of any square roots.

The operation counts for both the standard and square root free transformations are shown in table 2.2 The saving in using the square root free transformation clearly comes from the matrix multiply phase which effectively swamps the calculation of the rotation angles.

On the DAP with the standard method, the calculation of the rotation angles costs about the same as the multiply phase (for $n \leq 64$) which uses 2 matrix multiplies. Therefore we would not expect such great savings using square root free transformations as can be obtained on a serial computer. Indeed the calculation of the rotation angles for the square root free version takes longer than the standard case. The organisational overheads for the square root free DAP code are considerable and almost swamp out the saving of having one less matrix multiply. The rotation and multiply phases take about 1 ms and 0.7 ms respectively.

Table 2.2

Operations for one rotation

	Calculation of rotation angles				Matrix Multiply	
Standard	$\uparrow^2$	$\sqrt{\ }$	÷	+	*	+
	2	1	2	1	4N	2N
Square root free	$\uparrow^2$	*	÷	+	*	+
	2	4	3	1	2N	2N

3.1 Singular Value Decomposition

The Singular Value Decomposition (SVD) of a real m x n matrix A is defined as

$$A = U\Sigma V^t \tag{3.1}$$

where U is an m x n matrix, V is an n x n matrix

$U^tU = V^tV = I_n$, $\Sigma = \text{diag}(\sigma_1, \sigma_2 \ldots\ldots, \sigma_n)$.

The matrix U consists of the n orthogonal eigenvectors associated with the n largest eigenvalues of AA^T, and V consists of the eigenvectors of A^TA. The non-negative entries in Σ are the singular values.

3.2 Parallel Algorithm 1 for SVD via Jacobi rotations

Applying the parallel Jacobi method the plane rotations $\{p^{(k)}\}$ are performed on A^TA, and then A is transformed from the right.

$$A^{(k+1)} = A^{(k)}P^{(k)} \tag{3.2}$$

At each step (n/2) pairs of columns of A are simultaneously orthogonalised.

In the limit $A^{(k)} \underset{i\to\infty}{\to} U\Sigma$. The matrix V can be obtained by storing the product $\left(\prod_{i=1}^{i\to\infty} P^{(i)}\right)$, this can be computed simply by augmenting an (n x n) identity matrix with A before applying the transformations (3.2).

The parallel SVD 1 is defined as follows:

$A^{(1)} = A$

DO 1 SWEEPS = 1,2,......, UNTIL converged

DO 1 k = 1,2,....., (n-1)

$A^{(k+1)} = A^{(k)} P^{(k)}$

1 CONTINUE

where $\forall\ (p,q) \in Z_k$

$$\alpha^{(k)}(p,q) = 0.5 \tan^{-1}(2\underline{a}_i{}^t\underline{a}_j/(\underline{a}_i{}^t\underline{a}_i-\underline{a}_j{}^t\underline{a}_j))$$

$$P^{(k)}(p,q) = -P^{(k)}(q,p) = \sin(\alpha^{(k)}(p,q))$$

$$P^{(k)}(p,p) = P^{(k)}(q,q) = \cos(\alpha^{(k)}(p,q))$$

$$P^{(k)}(i,j) = 0\ \forall(i,j) \neq (p,q)$$

$$Z\{(p,q): 1 < p < q < n.\ p,q \text{ are columns of } A\}$$

$$\text{partition } Z,\ |Z_k| = n/2,\ k=1,2,...,(n-1)$$

$$V_k\ \{\alpha(p,q) : (p,q) \in Z_k\}$$

3.2.1 Choice of Ordering

The partitioning of Z into (n-1) independent sets is crucial in a parallel environment. There are many ways to pick these orders.† Implementation on the ICL DAP with a square grid topology revealed that the execution time strongly depends on the ordering chosen. A particular scheme, scheme Z, for which the organisational cost is less than 20% is defined as follows:-

Define logical matrices $D_{ij}^{(r)}$ $1 \leqslant r \leqslant n/2$, n even, i,j,k are integers

such that either $\{D_{ij}^{(r)}$ = .true. if $j=i+2r-1$ and $i+j+2r \equiv 1 \pmod 4\}$

or $\{D_{ij}^{(r)}$ = .true. if $j=i-2r+1+n$ and $i+j+2r \equiv (1+n) \pmod 4\}$

or $\{D_{ij}^{(r)}$ = .false.$\}$ (3.3a)

† For an interesting history, see Kirkmans School Girls Problem in Ball (1896)

<u>Define logical matrices $D_{ij}^{(r)}$ $(n/2 + 1) \leqslant r \leqslant n$</u>

such that either $\{D_{ij}^{(r)} = .\text{true}.\ \text{if}\ i+j \equiv 2(n+1-r)\,(\text{mod}\ n)\ \text{and}\ j>i\}$

$$\text{or}\ \{D_{ij}^{(r)} = .\text{false}.\} \qquad (3.3b)$$

Given the step number r, $1 \lesssim r \lesssim n$, the above formulae determines a logical matrix $D_{ij}^{(r)}$ which points to the elements to be annihilated at the rth step. Fig. 3.1(a) illustrates the ordering for a (16 x 16) example. Integer k denotes the simultaneous rotations carried out at the k'th step. For example when k=3, the pairs of columns

{(1.6),(2.13),(3.8),(4.15),(5.10),(6.1),(7.12),(8.3)
(9.14),(10.5),(11.16),(12.7),(13.2),(14.9),(15.4),(16.11)}

are simultaneously orthogonalised.

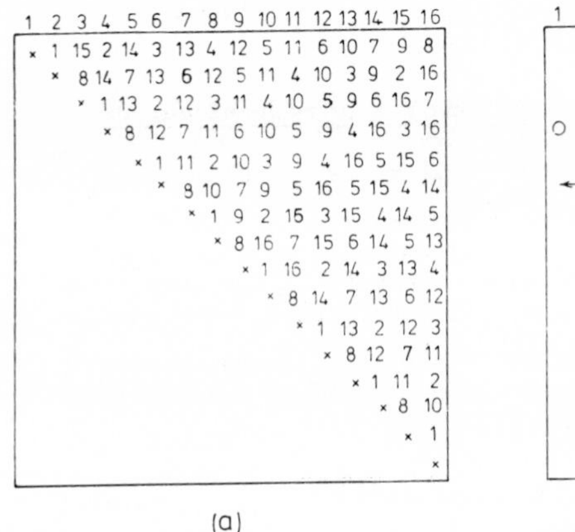

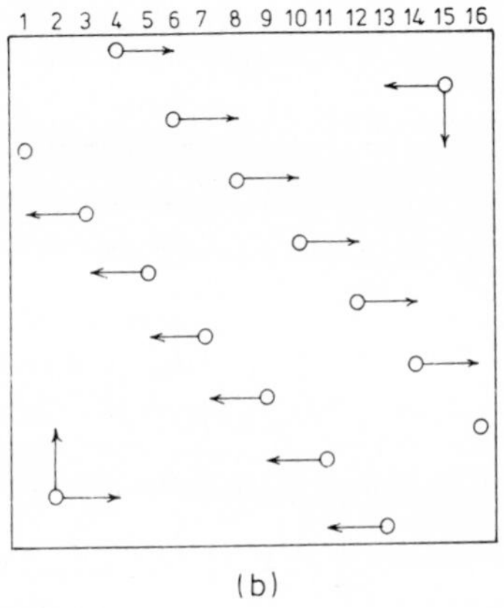

Fig. 3.1(a) Annihilation scheme Z applied to matrix of size 16,
(b) pattern generation

A novel method to reduce execution time is via a Mobile scheme - a technique designed specifically for parallel architecture, whereby after each parallel rotation the rows and columns of A are permuted so as to minimise the data movement overheads for subsequent rotations. A particular mobile scheme (MP1) for which the elements are always pivoted on the (2x2) blocks on the diagonal reduces the relocation cost to zero on any square grid architecture. For a detailed analysis on the mobile schemes the reader is referred to Modi and Pryce (1982).

3.2.2 Organisational Details

In this section we will discuss the implementation of SVD 1 utilising either a mobile or a stationary scheme (such as scheme Z). From (3.3a) and (3.3b) the parallelism is not immediately obvious. It seems ironical that endeavouring to exploit the parallelism should implicitly involve a serial description, which apparently disguises any 'pattern' that may exist. In this respect a diagrammatical description, such as that indicated in Fig. 3.1, appears to be more useful. In the organisational detail which follows we shall consider the following improvements:-

(i) utilising the parallel method to simultaneously find the position of the n/2 pairs of columns that are to be orthogonalised at each stage

(ii) simultaneously computing the set of angles $\{\alpha(p,q)\}$

(iii) developing an efficient scheme to compute the product $A^{(k)} * P^{(k)}$ and exploiting the pattern of non-zero elements in $P^{(k)}$.

In discussing the design features we will use the DAP FORTRAN language which provides a succinct notation.

(i) Determination of (n/2) pair of columns for orthogonalisation

The set Z_k gives the position of the (n/2) pairs of columns. Z_k is represented as a regular pattern using a logical array. The ability on the DAP to perform rapid logical operations means that it is not expensive to compute the patterns of pivot elements at any step. The set Z_{k+1} is obtained from Z_k by simple shift operations. For example, in Fig. 3.1(b), 'o's indicate the elements annihilated simultaneously at step 2. The arrows represent the movement needed to generate the pattern for step 3 from those of step 2.

(ii) Simultaneous computation of the set of angles

One stable method (Hammarling (1971), Rutishauser (1966)) of computing sine and cosine in order to avoid overflow and underflow is as follows†:-

$$\cot 2\,\alpha(p,q) = 0.5(\underline{a}_p^t\underline{a}_p - \underline{a}_q^t\underline{a}_q)/\underline{a}_p^t\underline{a}_q \tag{3.4}$$

$$\left.\begin{aligned} t_1 &= 1/|\cot 2\,\alpha\,(p,q)| + \sqrt{1 + \cot 2\,\alpha\,(p,q)} \\ \cos\alpha\,(p,q) &= 1/\sqrt{1 + t_1^{\,2}} \\ \sin\alpha\,(p,q) &= t_1 \cos\alpha(p,q) * \mathrm{sign}(\cot 2\,\alpha(p,q)) \end{aligned}\right\} \tag{3.5}$$

† superscript k has been omitted for clarity

The formulae (3.4) and (3.5) transform into parallel using arithmetic in vector mode, and a detailed descrition can be found in Modi (1982).

(iii) Schematic matrix multiplier

Matrix A is partitioned into blocks of (64 x 64) units $\{A_1, A_2 \ldots A_\ell\}$. Partitioning of matrix A, of dimension $(64\ell) \times 64$, $\ell \geq 2$, is indicated in Fig. 3.2.

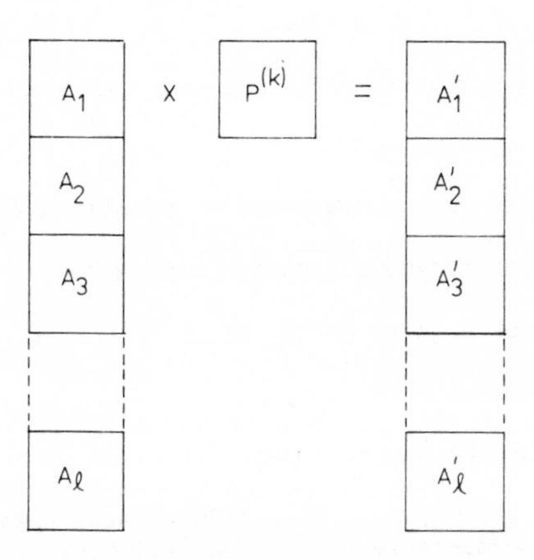

Fig. 3.2 Partition of $(64\ell) \times 64$ size matrix

Clearly: $A_1' = A_1 P^{(k)}$, $A_2' = A_2 P^{(k)}$, ...$A_\ell' = A_\ell P^{(k)}$.

Each product $A_i P^{(k)}$ consists of a (64 x 64) dense matrix A_i and a (64 x 64) orthogonal matrix $P^{(k)}$ (which has a regular pattern of non-zero elements). The following multiplication algorithm exploits the pattern of non-zeros in $P^{(k)}$ and does the multiplication in minimum arithmetic.

If $P^{(k)}$ is a matrix as defined in §3.2 with partition Z_k then we define permutation α, in cyclic notation as

$$\alpha = \prod_{(p,q) \in Z_k} (p,q)$$

We introduce the notation for column permutation of a matrix with α.

$$\underline{Z}(i,j) = Z(i,\alpha(j))$$

With this notation the product $X^{(k+1)} = A_i P^{(k)}$ can be expressed as follows:

$$X^{(k+1)} = C^{(k)T}.A_i + S^{(k)T}.A_{-i} \tag{3.6}$$

where $C^{(k)}(i,j) = P^{(k)}(i,i) \;\forall\; i,j$

$S^{(k)}(i,1) = P^{(k)}(i,\ell) \;\forall\; i \neq \ell, \;\{1 \leq \ell \leq n \;;\; P^{(k)}(i,\ell) \neq 0\}$

$S^{(k)}(i,j) = S^{(k)}(i,1) \;\forall\; i,j$

(.) indicates element by element multiplication

In DAP FORTRAN (3.6) may be coded as:

```
REAL A(,),SINEVECTOR(),COSINEVECTOR(),PERMUTEDA(,)

A=MATR(SINEVECTOR)*PERMUTEDA+MATR(COSINEVECTOR)*A
```

where PERMUTEDA is a column permutation of A.

3.2.3 Numerical Properties

The SVD1 is based on the use of independent sets of rotations. The special cyclic orderings for which the convergence has been established (Forsythe and Henrici (1960), Wilkinson (1962)) are, however, not suitable for parallel computation, where General Cyclic Orderings are utilised. The convergence properties of the special cyclic ordering do not appear to carry over to the General Cyclic Ordering. However, with angle restriction convergence for General Cyclic Ordering has been established (Brodlie and Powell (1975)).

For parallel computation threshold strategy may be applied as indicated in Theorem 3.1.

Theorem 3.1

For a parallel Jacobi rotation where the non-interacting plane rotations R_1 ....., $R_m (m \in Z)$ are applied simultaneously on the pivots $A_{p_r q_r}$ (r=1,2,...,m) of a real symmetric matrix $A^{(k)}$, it is sufficient to guarantee convergence if one of the ${A^{(k)}_{p_r q_r}}^2 > cS^{(k)}(A)$, where $S^{(k)}(A) = \frac{1}{2} \sum_{i \neq j} {A^{(k)}_{ij}}^2$, $c=1/\frac{n}{2}(n-1)$ and n>2 is the order of the matrix $A^{(k)}$.

Proof

Let $S^{(k)}(A) = \frac{1}{2} \sum_{i \neq j} {A^{(k)}_{ij}}^2$, $\quad k = 1,2,3,......$

then for non-interacting rotations $(p_1q_1)....(p_mq_m)$

$$S^{(k)}(R_{p_m q_m} \cdots R_{p_1 q_1} A^{(k)} R^t_{p_1 q_1} \cdots R^t_{p_m q_m}) = S^{(k)}(A) - \sum_{r=1}^{m} A^{(k)^2}_{p_r q_r} \tag{3.7}$$

$$= m^{(k)} S^{(k)}(A)$$

where $$m^{(k)} = (1 - \sum_{r=1}^{m} A^{(k)^2}_{p_r q_r} / S^{(k)}(A))$$

and if one of the $A^{(k)^2}_{p_r q_r} > c\, S^{(k)}(A)$

then $m^{(k)} < (1-c)$

iterate (3.7)

$$S^{(k+1)}(A) = m^{(k)} . m^{(k-1)} . m^{(k-2)} \ldots m^{(1)} S^{(1)} < (\prod_{i=1}^{k} m^{(i)}) S^{(1)}$$

and $$\prod_{i=1}^{k} (m^{(i)}) < (1-c)^{(k)}$$

therefore $S^{(k+1)} \to 0$ as $k \to \infty$.

Forsythe and Henrici (1960) have shown that the statements,

$$\lim_{k\to\infty} S^{(k)}(A) = 0$$

and $$\lim_{k\to\infty} A^{(k)} = D, \text{ a diagonal matrix}$$

are equivalent. Therefore the matrix $A^{(k)}$ converges to a fixed diagonal matrix D as $k\to\infty$. This completes the proof.

(ii) Error Analysis for the Parallel Algorithm

The method is essentially the Jacobi method applied on $A^T A$. A detailed error analysis for the parallel Jacobi method has been carried out by Modi (1982). After one sweep, if $N=(n-1)$, $x = 6.2^{-t}$,

$$\left.\begin{aligned} &\|\bar{A}^{(N)} - A^{(O)}P^{(1)}P^{(2)} \;\ldots\ldots\; P^{(N)}\| \leqslant x\,(n-1)(1+x)^{(n-2)}\|A^{(O)}\| \\ &\|\bar{V}^{(N)} - P^{(1)}P^{(2)} \;\ldots\ldots\; P^{(N)}\| \leqslant x\,n^{3/2}\,(1+x)^{(n-2)} \end{aligned}\right\} \tag{3.8}$$

The bounds in (3.8) are for (nxn) matrix A, t is the machine accuracy, $\bar{A}^{(N)}, \bar{V}^{(N)}$ are the computed values of $A=U\Sigma$ and V respectively. The bounds are a slight improvement on the special cyclic ordering.

In Table 3.1 comparison is made with a similar algorithm implemented on the Illiac IV.

Table 3.1

Comparison of results for a matrix of order 64, on the DAP and Illiac

	SWEEPS	Time(sec)	Order of the algorithm (m=n)	Algorithm
SVD1 ON DAP	6	1.5	$k\left[\frac{n}{64}\right] \times n$	Jacobi type
Illiac IV	9	4.56	$k\left[\frac{n}{64}\right] \times n^2$	Jacobi type
NAG	-	7.3	$2n^3$	Householder + QR algorithm

It is seen that for higher order matrices SVD1 on the DAP should perform better than Illiac IV. This is because Illiac IV only does vector operations of order n in parallel.

3.3 Parallel Algorithm 2 for SVD, via Givens rotations

In SVD1 the iterations were performed on the entire plane of A. In this section we develop a parallel algorithm which reduces it to a smaller size.

The SVD2 for a rectangular matrix A, mxn, m>n, is defined in two steps.

Step 1 Perform a QU factorization of A where Q_1 is an mxm orthogonal matrix so that $Q_1^t Q_1 = I_m$, and $\tilde{U}$ is an nxn upper triangular matrix,

$$A=Q_1 \begin{bmatrix} \tilde{U} \\ O \end{bmatrix}$$

The parallel QU factorization developed in §2 is used here.

<u>Step 2</u> Find the SVD of $\tilde{U}$

$$\tilde{U} = Q_2 \Sigma P^t, \text{ where } \Sigma = \begin{bmatrix} \sigma_1 & & & O \\ & \sigma_2 & & \\ & & \ddots & \\ O & & & \sigma_n \end{bmatrix}$$

Q_2, Σ, P^t are all (nxn) matrices

$$\sigma_1 \geq \sigma_2 \geq \ldots\ldots \geq \sigma_n \geq 0$$

Then SVD of A is

$$A=Q_1 \begin{bmatrix} Q_2 & O \\ O & I_{m-n} \end{bmatrix} \begin{bmatrix} \Sigma \\ O \end{bmatrix} P^t$$

The organisational details for step 1 are discussed in §2 and for step 2 are discussed in §3.2.2.

3.3.1 Numerical Properties

Error bounds for step 1 and step 2 are considered next.

<u>Step 1</u> The error analysis of the parallel QU factorization has been carried out by Gentleman (1975)

After (m+n-2) steps

$$\|A^{(W)} - Q^{(W)} Q^{(W-1)} \ldots\ldots Q^{(1)} A^{(O)}\| < x \,.(m+n-2).(1+x)^{m+n-1} \|A^{(O)}\| \qquad (3.9)$$

where $W=m+n-2$, $A^{(W)}=\tilde{U}$ and $x=6.2^{-t}$

<u>Step 2</u> Error bounds for step 2 are given by equation (3.8). (3.8) and (3.9) indicate that both steps are numerically stable.

3.4 Timing Comparison of SVD1, SVD2 and the NAG routine

Fig. 3.3 illustrates the relative efficiency of SVD1, SVD2 and the NAG routine, which is a different algorithm (Golub and Reinsch (1970)). It is seen that SVD2 is superior to the others.

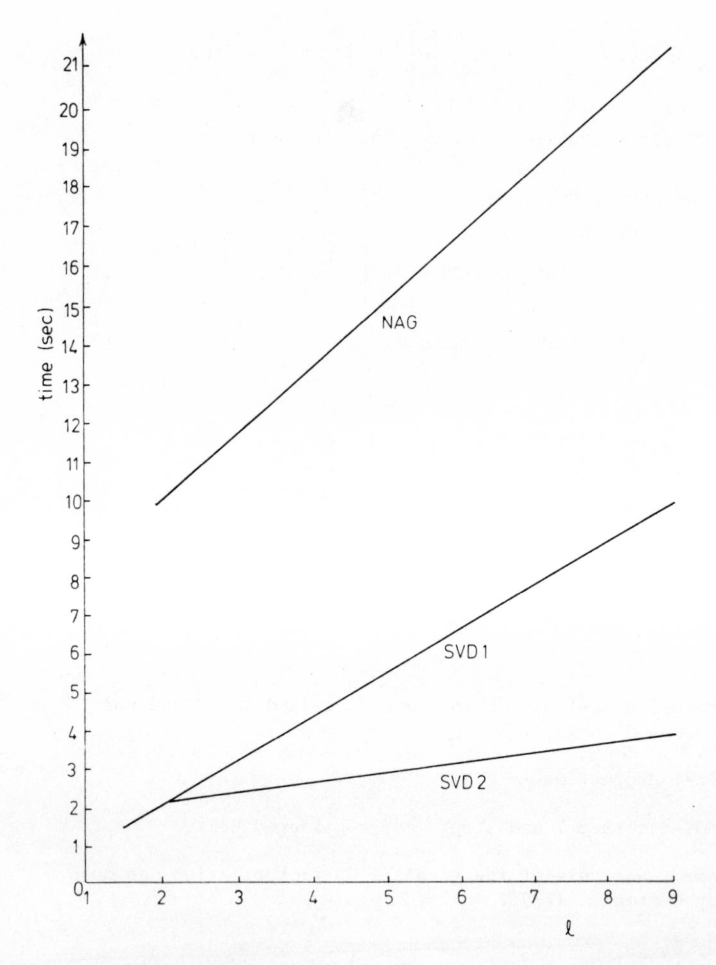

Fig. 3.3 Timing comparison for SVD1/SVD2/NAG for 64ℓx64 size matrices

4. CONCLUSIONS AND EXTENSIONS

The DAP architecture brings into consideration orthogonal techniques where several non-interacting rotations can be used as the basis of many algorithms in numerical analysis. Extension to Generalised Eigenvalue problem has been considered by Modi (1982).

The implementation of algorithms is much more crucial in parallel environment.

The QU factorization and the SVD algorithms, although designed for the DAP, can be applied to other similar SIMD architectures with little modification.

The QU factorization can also be performed using elementary Householder reflections. Initial experiments indicate that the Householder scheme may be more amenable to the DAP. (See Bowgen and Modi (1982)).

REFERENCES

Ball, W.W.R. (1896) Mathematical recreations and essays, Macmillan and Co., London.

Bowgen, G.S.J. and Modi, J.J. (1982) Householder QU factorization on the DAP, Internal Report, DAPSU, QMC, London E1 4NS.

Brodlie, K.W. and Powell, M.J.D. (1975) On the convergence of cyclic Jacobi methods, *J. Inst. Math. Appl.*, **15**, pp. 279-287.

Clarke, M.R.B. and Modi, J.J. (1982) A note on Givens ordering, Internal Report, Dept. of Computer Sci. and Statistics, Queen Mary College, London E1 4NS.

Forsythe, G.E. and Henrici, P. (1960) The Cyclic Jacobi method for Computing the principal values of a complex matrix, *Trans. Amer. Soc.*, **94**, pp. 1-23.

Gentleman, W.M. (1975) Error Analysis of QR decomposition by Givens transformation, *Lin. Alg. Appl.*, **10**, pp. 189-197.

Gentleman, W.M. (1973) Least squares computations by Givens transformations without square roots, *J. Inst. Math. Appl.*, **12**, pp. 329-336.

Golub, G.H. and Reinsch, C. (1970) Singular value decomposition and least squares solutions, *Numer. Math.*, **14**, pp. 403-420.

Golub, G.H. and Kahan, W. (1965) Calculating the singular values and pseudo-inverse of a matrix, *J. SIAM ser. B-Numer. Anal.* **2**, pp. 205-224.

Hammarling, S.A. (1974) A note on modification to the Givens plane of a rotation, *J. Inst. Math. Appl.*, **13**, pp. 215-218.

Henrici, P. (1958) On the speed of convergence of cyclic and quasi-cyclic Jacobi methods for computing eigenvalues of Hermitian matrices, *J. SIAM*, **6**, pp. 144-162.

Hestenes, M.R. (1958) Inversion of matrices by biorthogonalization and related results, *J. SIAM*, **6**, pp. 51-90.

Luk, F.T. (1980) Computing the singular value decomposition on Illiac IV, *ACM Tran. Math. Software*, **6**, no. 4, pp. 514-539.

Modi, J.J. (1982) Jacobi methods for eigenvalue and related problems in a parallel computing environment, Ph.D Thesis, University of London.

Modi, J.J. and Parkinson, D. (1982) Study of Jacobi methods for eigenvalue and singular value decomposition on DAP, Conf. Proc., Computer Physics Communications, pp. 317-320.

Modi, J.J. and Pryce, J.D. (1982) Efficient implementation of Jacobi's method on the DAP, submitted for publication.

Ruitishauser, H. (1966) The Jacobi method for real symmetric matrices, *Numer. Math.*, **9**, pp. 1-10.

Sameh, A.H. and Kuck, D.J. (1978) On parallel linear system solvers, *JACM,* **25**, pp. 81-91.

Wilkinson, J.H. (1977) Some recent advances in numerical linear algebra, in The state of the Art in Numerical Analysis, (D. Jacobs, ed.), Academic Press, pp. 3-53.

Wilkinson, J.H. (1962) Note on the quadratic convergence of the cyclic Jacobi process, *Numer. Math.*, **4**, pp. 296-300.

IMPLEMENTATION OF A PARALLEL (SIMD) MODIFIED NEWTON ALGORITHM ON THE ICL DAP

K.D. Patel

(Numerical Optimisation Centre, Hatfield Polytechnic)

ABSTRACT

A parallel (SIMD) version of Modified Newton algorithm was implemented on the ICL DAP system. A general description of the algorithm followed by a discussion on where and how parallelism on the DAP was utilised is outlined. A set of results comparing the parallel algorithm with a number of sequential algorithms is presented and the 'speed-up' factors calculated.

The difficulties encountered during the implementation are also highlighted.

1. INTRODUCTION

We are interested in solving the unconstrained optimisation problem

$$\min \qquad f(\underline{x}), \qquad \underline{x} \in \mathbb{R}^n$$

possibly with simple upper and lower bounds

$$l_i \leqslant x_i \leqslant u_i$$

using the currently available parallel systems. The parallel system we used was the ICL Distributed Array Processor (DAP) at Queen Mary College, London.

The ICL DAP is classified as a Single Instruction Multiple Data (SIMD) machine. The ICL DAP comprises 4096 processing elements arranged in a (64 x 64) matrix form, all processors obey a single instruction stream broadcast by a Master Control Unit. A detailed description of the ICL DAP can be found elsewhere in the proceedings. The class of problems we are interested in solving are small expensive problems. The reader is referred to Dixon (1981) for the classification of problems and possible interaction of them with the currently available parallel processing machines.

Small Expensive Problems (Modified Newton technique)

Consider the interaction of small expensive problems with the SIMD ICL DAP architecture. On a sequential system the modified Newton techniques for solving unconstrained optimisation problems is one of the most efficient for n in the range 2 to 5. By introducing parallelism into the Modified Newton technique, suitable for implementation on the ICL DAP, we hope to increase the range for which Modified Newton technique is most efficient. First we outline the factors involved in the sequential Modified Newton technique and then consider ways in which we can introduce parallelism in the algorithm.

Nearly all efficient methods for sequentially solving unconstrained optimisation problems require the gradient vector $\underline{g}$ to be evaluated at $\underline{x}^{(k)}$. For some very expensive industrial problems, which may include simulation, the calculation of $\underline{g}$ can be very expensive. We can use estimated values of $\underline{g}$, obtained by some finite differences scheme, and this would be an ideal calculation for the DAP. There are efficient Sequential Variable Metric and Conjugate Gradient algorithms which use central difference scheme to compute $\underline{g}$. So this may be acceptable on a parallel machine. Refer to Fletcher (1982) for the theory of Newton, Variable Metric and Conjugate Gradient algorithms.

The Modified Newton approach is classified as a second derivative method, i.e. not only do we need to calculate $\underline{g}$, but we also need to calculate the Hessian matrix G. The elements of G are given by

$$G_{ij} = \frac{\partial^2 f}{\partial x_i \partial x_j}$$

Again we could approximate G by some finite differences scheme.

The overhead in solving

$$(G + \mu I)\underline{p} = -\underline{g}$$

is in determining the search direction $\underline{p}$ on the DAP. μ is always chosen such that the modified matrix $(G + \mu I)$ is positive definite.

The final stage in the Modified Newton algorithm is to select the step size α. This is often done by a line search or by a search along a preselected curve. For the DAP implementation we considered three possible searches.

(a) '1-D' search (line search).

(b) '2-D' search } appropriate choice of planes.

(c) '4-D' search }

The '4-D' subspace search seems to be the most attractive proposition. For problems with $4 < n \leq 64$, the parallel algorithm based on this idea will be considerably more efficient and also reduce the number of Newton iterations. In the next section we describe the parallel Modified Newton algorithm which we implemented on the DAP.

2. PARALLEL (SIMD) MODIFIED NEWTON ALGORITHM

The parallel Modified Newton algorithm consists of the following steps:

Step 1: Initial guess $\underline{x}^{(0)}$; h, the step-length and $\in$, the tolerance

Step 2 (a): Calculate $f(\underline{x} \pm h\,\underline{a}_i \pm h\underline{a}_j)$ all $j > i$, where a_i is a unit vector along the ith axis. This function evaluation is a parallel calculation.

Step 2(b): Calculate gradient vector $\underline{g}$ and the Hessian matrix G by finite differences scheme.

Step 3: Stop if $\max|g_i| < \in$.

Step 4: Solve a set of linear simultaneous equations

$$(G + \mu I)\underline{p} = -\underline{g}.$$

Step 5: We considered three possible searches:

(i) one-dimensional search (line search)

$$X^{(k+1)} = \underline{x}^{(k)} + (A1)\alpha_1\underline{p}^{(k)}$$

(ii) two-dimensional search

$$X^{(k+1)} = \underline{x}^{(k)} + (A1)\alpha_1\underline{p}^{(k)} + (A2)\alpha_2\underline{g}^{(k)}.$$

(iii) four-dimensional search

$$X^{(k+1)} = \underline{x}^{(k)} + (A1)\alpha_1\underline{p}^{(k)} + (A2)\alpha_2\underline{g}^{(k)} + (A3)\alpha_3\underline{d}_3^{(k)} + (A4)\alpha_4\underline{d}_4^{(k)}$$

where scalars α_1, α_2, α_3 and α_4 and (64 x 64) matrices A1, A2, A3 and A4 are specified later in this section.

Step 6: $$x^{(k+1)} = \text{Arg} \min_{i=1,\dots 4096} f({}^{i}X^{(k+1)})$$

- minimising over a preselected grid.

Step 7: Return to 2 with k = k+1.

Some more details on some individual steps may ease a more general discussion of the parallel algorithm.

Step 2(a): The function values at the points $\underline{x}^{(k)} \pm h\underline{a}_i \pm h\underline{a}_j$ are stored in the 64 x 64 array F. For a (4 x 4) DAP and a 4-dimensional problem (n = 4).

$$F = \begin{bmatrix} f(\underline{x}+ha_1) & f(\underline{x}+ha_1+ha_2) & f(\underline{x}+ha_1+ha_3) & f(\underline{x}+ha_1+ha_4) \\ f(\underline{x}+ha_2) & f(\underline{x}-ha_2) & f(\underline{x}+ha_2+ha_3) & f(\underline{x}+ha_2+ha_4) \\ f(\underline{x}+ha_3) & f(\underline{x}-ha_3) & & f(\underline{x}+ha_3+ha_4) \\ f(\underline{x}+ha_4) & f(\underline{x}-ha_4) & f(\underline{x}-ha_1) & f(\underline{x}) \end{bmatrix}$$

Step 2(b): The following finite-difference scheme were used to calculate the gradient vector $\underline{g}$ and the Hessian matrix G

$$g_i = [f(\underline{x}+ha_i) - f(\underline{x}-ha_i)]/2h$$

$$G_{ij} = [f(\underline{x}+ha_i+ha_j) - f(\underline{x}+ha_i) - f(\underline{x}+ha_j) + f(\underline{x})]/h^2$$

$$i \neq j$$

$$G_{ii} = [f(\underline{x}+ha_i) - 2f(\underline{x}) + f(\underline{x}-ha_i)]/h^2$$

Step 4: We used DAP library subroutine FO4GJNLE64 to solve the set of linear simultaneous equations

$$G\underline{p} = -\underline{g}.$$

Step 5:

(i) 4-dimensional grid search.

The iterative step given by

$$X^{(k+1)} = \underline{x}^{(k)} + (A1)\alpha_1\underline{p}^{(k)} + (A2)\alpha_2\underline{g}^{(k)}$$

$$+ (A3)\alpha_3\underline{d}_3^{(k)} + (A4)\alpha_4\underline{d}_4^{(k)},$$

generates 4096 points

where $\alpha_1 = \alpha_1^{\text{initial}}$,

$$\alpha_2 = -\alpha_1 * \frac{g^T g}{g^T G g},$$

$$\alpha_3 = -\alpha_1 * \frac{d_3^T g}{d_3^T G d_3},$$

$$\alpha_4 = -\alpha_1 * \frac{d_4^T g}{d_4^T G d_4}, \qquad (2.1)$$

$$\underline{d}_3^{(k)} = \underline{x}^{(k)} - \underline{x}^{(k-1)},$$

$$\underline{d}_4^{(k)} = \underline{x}^{(k-1)} - \underline{x}^{(k-2)}, \tag{2.1}$$

A1, A2, A3 and A4 are (64 x 64) matrices given by

$$A1_{\text{ith row}} = [-2,-2,-2,-2,-2,-2,-2,-2\vdots -1,-1,-1,-1,-1,-1,-1,-1\vdots 0,\ldots 0\vdots 1,\ldots\ldots 1\vdots 2,\ldots\ldots 2\vdots 3,\ldots\ldots 3\vdots 4,\ldots\ldots 4\vdots 5,\ldots\ldots 5]. \tag{2.2}$$

$$A2_{\text{ith row}} = [-2,-1,0,1,2,3,4,5\vdots -2,-1,0,1,2,3,4,5\vdots \ldots\ldots\vdots -2,-1,0,1,2,3,4,5] \tag{2.3}$$

$$A3_{\text{ith col}} = [-3,-3,-3,-3,-3,-3,-3,-3,\vdots -2,-2,-2,-2,-2,-2,-2,-2\vdots 1,\ldots -1\vdots 0,\ldots 0\vdots 1,\ldots 1\vdots 2,\ldots 2\vdots 3,\ldots\ldots 3\vdots 4,\ldots\ldots 4] \tag{2.4}$$

$$A4_{\text{ith col}} = [-3,-2,-1,0,1,2,3,4\vdots -3,-2,-1,0,1,2,3,4\vdots \ldots\ldots\vdots -3,-2,-1,0,1,2,3,4]. \tag{2.5}$$

The 4096 points generated are ${}^{i}X$, $i=1,2,\ldots.4096$

if $\min_{i=1,\ldots.4096} f({}^{i}X^{(k+1)}) = f(\underline{x}^{(k)})$

then set $\alpha_1 = \frac{\alpha_1}{10}$ and recompute search step.

As soon as

$$\min_{i = 1,\ldots 4096} f({}^{i}X^{(k+1)}) < f(\underline{x}^{(k)})$$

then reset $\alpha_1 = \alpha_1^{\text{initial}}$.

(ii) 2-dimensional grid search

The iterative step is

$$X^{(k+1)} = \underline{x}^{(k)} + (A1)\alpha_1\underline{p}^{(k)} + (A2)\alpha_2\underline{g}^{(k)}$$

where matrices A1 and A2 are given by:

(a) (2.2) and (2.3) respectively,
(We call this VERSION A),

(b) $A1_{\text{jth col}} = [-16,-15.5,-15,-14.5,-14, \ldots .14.5,15,15.5]$

$A2_{\text{ith row}} = [-16,-15.5,-15,-14.5,-14, \ldots .14.5,15,15.5]$
(We call this VERSION B).

The iterative step will generate 4096 points; in fact only 64 distinct points for VERSION A, the rest are identical to these. For VERSION B, the iterative step will generate 4096 distinct points. The α's are given by (2.1).

For the 2-dimensional grid search, we considered the plane defined by $\underline{p}^{(k)}$ and $\underline{g}^{(k)}$, as this plane contains directions which can provide recovery from points at which G is singular.

(iii) 1-dimensional grid search.

The iterative step is

$$x^{(k+1)} = \underline{x}^{(k)} + (A1)\alpha_1\underline{p}^{(k)}$$

where the matrix A1 is given by

(a) (2.2) [VERSION A],

(b) A1 has the values -102(0.05)102.75 in long vector order [VERSION B].

The iterative step will generate 4096 points; in fact only 8 distinct points for VERSION A, the rest are identical to these.

This is basically a line search, with 8 points on the line for VERSION A and 4096 points on the line for VERSION B.

The strong feature of the parallel algorithm is the function evaluation (steps 2 to 6). The DAP can perform 4096 function evaluations in parallel. It would be naive to assume that since the DAP performs 4096 function evaluations in parallel, the processing time would be of the order 4096 faster than the sequential calculation. In fact the arithmetic operations in the processing elements of the DAP are slower than the fast sequential machines (e.g. CDC 7600); the bit serial nature of each of the 4096 processing elements in the DAP accounts for the slowness in the arithmetic operations. Let the ratios of the speeds be τ. To get a rough idea of the value of τ, we obtained processing times for 4096 function evaluations of five 64-dimensional test problems (specified in the next section) on ICL DAP, ICL 2980 and DEC 1091. The processing times are displayed in table (2.1).

Table (2.1)

The processing times for 4096 function evaluations (time in seconds)

Functions	Parallel evaluation DAP	Sequential evaluation	
		DEC 1091	ICL 2980
Quadratic	0.045832	2.601	1.00928
Rosenbrock	0.109032	3.963	1.315296
Powell	0.067200	4.453	1.563352
Box (M)	11.62174	189.779	93.995920
Trigonometric	0.351760	51.267	15.554040
Average ratio	1	20.7	9.3

The subroutine for 4096 function evaluations on the DAP, for Rosenbrock function, is as follows.

```
REAL  MATRIX  FUNCTION  FUNCT(X)

REAL  X(,,64)

DO  10  I = 1,64,2

10  FUNCT = FUNCT+ 100.0* (X(,,I)**2 - X(,,I+1))**2 + (1.0 - X(,,I))**2

RETURN

END
```

where X is a (64 x 64 x 64) array.

Element X(I,J,K) contains the numerical value of the Kth co-ordinate of the point whose position on a (64 x 64) grid is at (I,J).

3. EXPERIMENTAL RESULTS

Performance Measurement

On the question of performance measurement, the 'speed-up' and the 'efficiency' ratios are meaningless concepts for the DAP type architecture. The criteria we have used for measuring the performance of the parallel algorithm is to compare the processing times obtained on the DAP in relation to a sequential system.

Note of caution: Often the optimum DAP algorithm differs from the optimum sequential algorithm and this complicates comparisons.

The sequential system we used was the DEC 1091 at Hatfield Polytechnic. The codes used for the sequential system were:

(a) Standard Newton-Raphson method from the NAG Library, routine E04EBF[3], the nearest equivalent sequential algorithm.

(b) Variable Metric algorithm, the recommended algorithm, for a sequential system, for a 64 dimensional problem. We selected the Numerical Optimisation Centre's OPTIMA Library program OPVM[4], an implementation of the Broyden-Fletcher-Shanno variable metric approach to unconstrained optimisation.

(c) Noticing that the structure of some of the test functions was symmetric and noting that this symmetry would facour a Conjugate Gradient approach, the Harwell Library routine VA14A[5] was also used.

Choice of Step-Length, h

One of the main difficulties encountered during the implementation was the choice of h, the step-length, used in the finite difference schemes for approximating $\underline{g}$ and G.

To illustrate this, consider the 64-dimensional Rosenbrock function. $\underline{g}$ and G are approximated using the finite difference schemes stated in Section 2.

The expressions for $g_1, g_2, G_{11}, G_{22}, G_{12}$ are:

$$\begin{aligned} g_1 &= 400x_1(x_1^2+h^2) - 400x_1x_2 - 2 + 2x_1, \\ g_2 &= 200(x_2-x_1^2), \\ G_{11} &= 100[12x_1^2+2h^2-4x_2] + 2, \\ G_{22} &= 200, \\ G_{12} &= G_{21} = -400x_1 - 200h. \end{aligned} \tag{3.1}$$

Substitute for $\underline{x}^* = (1,1,\ldots\ldots 1)^T$, local minimum, in (3.1) to obtain

$$\underline{g} = (400h^2,\ 0, 400h^2,\ 0,\ldots\ldots)^T.$$

The stopping criterion we have used is

$$\max|g_i| < \epsilon.$$

Now $\max|g_i| = 400h^2 \quad (=g_1)$

i.e. we need $400h^2 < \epsilon$.

Tabulate g_1 $(=400h^2)$ for four different values of h.

h	g_1
0.1	4.0
0.01	0.04
0.001	0.0004
0.0001	0.000004

If $\in = 0.001$, then only h = 0.001 and h = 0.0001 will satisfy the stop criteria.

Now take $\underline{x} = (-1.2,1.0,-1.2,1.0,.....,-1.2,1.0)^T$, used as a starting vector. Table 3.1 displays the values of g_1, g_2, G_{11}, G_{22} and G_{12} for different values of h.

For h =0.0001, the DAP calculated values differ greatly from the actual values for G_{11}, G_{22}, G_{12}. This discrepancy is due to rounding errors in the DAP calculations. In the DAP-FORTRAN code all the real variables XX, G and HESS are declared as REAL*4 (32 bits of DAP store is used to represent a real value).

XX - (64x64x64) array. Initially stores the values of $\underline{x} \pm h\underline{a}_i \pm h\underline{a}_j$, and then the values of the co-ordinates of 4096 points. This occupies half of the available DAP store.

G - 64 element vector. Stores the gradient.

HESS - (64x64) array. Stores the Hessian.

The potential inaccuracy in the mantissa of a real value is of the order $0.5*10^{-6}$, i.e. the real value is accurate to approximately 7 decimal digits. To obtain a fairly good approximation of the Hessian (increase the accuracy of a real value to approximately 16 decimal digits), we need to declare XX(,,64), G() and HESS(,) as REAL*8. But this means that the real values of XX will occupy the whole DAP store.

With XX(,,64), G() and HESS(,) declared as REAL*4 variables, the next obvious step would be to increase h, bearing in mind that a 'large' value of h would not satisfy the stop criteria at the exact solution. A suitable choice of h was 0.001 with $\in = 0.001$.

Numerical Results

The five test problems were run on the

(i) DAP, using the parallel algorithm

(ii) DEC 1091 using

(a) Modified Newton-Raphson NAG routine, E04EBF [3]

(b) Variable Metric, NOC OPVM [4]

(c) Conjugate Gradient, Harwell VA14A [5].

Table 3.1

	g_1	g_2	G_{11}	G_{22}	G_{12}
h=0.1					
DAP calculation	-220.4005	-88.00006	1332.028	200.0018	460.0007
exact value	-220.4	-88.0	1332.0	200.0	460.0
h=0.01					
DAP calculation	-215.64960	-88.00124	1330.28	200.0427	478.05760
exact value	-215.648	-88.0	1330.02	200.0	478.0
h=0.001					
DAP calculation	-215.7211	-88.03558	1499.176	198.36430	488.28130
exact value	-215.60048	-88.0	1330.0002	200.0	479.8
h=0.0001					
DAP calculation	-215.50650	-88.19580	10776.52	0.0	1525.87900
exact value	-215.6000048	-88.0	1330.000002	200.0	479.98

exact values are calculated from the analytic expressions (3.1)

We used approximate values of $\underline{g}$, gradient vector and G, the Hessian, obtained using finite difference schemes described in Section 2. For all the test problems, we used two sets of starting points, a symmetric and a non-symmetric set.

The numerical results are displayed in tables (A.1) to (A.8).

Table (A.1) displays the processing times for the sequential codes.
Table (A.2) displays the DAP processing times for '1-D' search (VERSION A).
Table (A.3) displays the DAP processing times for '1-D' search (VERSION B).
Table (A.4) displays the DAP processing times for '2-D' search (VERSION A).
Table (A.5) displays the DAP processing times for '2-D' search (VERSION B).
Table (A.6) displays the DAP processing times for '4-D' search.
The 'speed-up' ratio, for performance measurement, is displayed in tables (A.7) and (A.8).

It will be seen from the numerical results that the parallel algorithm consistently outperformed the Newton-Raphson and Variable Metric sequential algorithms. In fact the DAP performed extremely well compared with the sequential Newton-Raphson algorithm.

Note of caution: Although the sequential Newton-Raphson is the nearest equivalent sequential algorithm to the parallel one, it is rarely used to solve a 64-dimensional problem.

For problems with special symmetry in the objective function, the parallel algorithm did perform better than the Conjugate-Gradient sequential algorithm, but not by a large factor (in terms of 'speed-up' ratio). But for the non-symmetric test function (Trignometric function) the parallel algorithm did perform a lot better than the sequential Conjugate-Gradient algorithm.

Consider the '1-D' search algorithm, VERSION A performed better than VERSION B. For the '2-D' search algorithm, VERSION B performed better than VERSION A. Looking at the performance ratio and the number of iterations, there is not much to choose between the '2-D' search and the '4-D' search; for some problems the '2-D' search performed better than the '4-D' search and for other problems the '4-D' search performed better than the '2-D' search.

ACKNOWLEDGEMENTS

The author thankfully acknowledges the support of the SERC grant No. GR/B/4665/5; and the use of the DAP machine at Queen Mary College, London.

REFERENCES

Dixon, L.C.W. (1981) The place of parallel Computation in Numerical Optimisation I, The Local Problem. Proceedings of the EEC/CNR Summer School, Design of numerical algorithms for parallel processing, Bergamo University. 1981

Fletcher, R. (1980) Practical methods of optimisation. Unconstrained optimisation, John Wiley.

NAG Manuals (MARK 8).

OPTIMA Manual. Numerical Optimisation Centre. Hatfield Polytehcnic. Issue No. 4 (1982).

HARWELL Subroutine library (1981).

APPENDIX

The following five 64-dimensional test problems were used

(a) Quadratic function

$$f(\underline{x}) = \sum_{i=1}^{32} 100x_{2i}^2 + (1-x_{2i-1})^2.$$

(b) Rosenbrock function

$$f(\underline{x}) = \sum_{i=1}^{32} 100(x_{2i-1}^2 - x_{2i})^2 + (1-x_{2i-1})^2.$$

(c) Powell function

$$f(\underline{x}) = \sum_{i=1}^{16} [(x_{4i-3} + 10x_{4i-2})^2 + 5(x_{4i-2} - x_{4i})^2 + (x_{4i-2} - 2x_{4i-1})^4 + 10(x_{4i-3} - x_{4i})^4].$$

(d) Box (M) function

$$f(x) = \sum_{j=1}^{32} \sum_{i=1}^{10} [\exp(-x_{2j-1} t_i) - 5\exp(-x_{2j} t_i) - \exp(-t_i) + 5\exp(-10t_i)]^2,$$

$$t_i = \frac{i}{10}, \; i=1,2,\ldots.10.$$

(e) Trigonometric function

$$f(\underline{x}) = \sum_{i=1}^{64} [64 + i - \sum_{j=1}^{64} (A_{ij} \operatorname{Sin}(x_j) + B_{ij} \operatorname{Cos}(x_j))]^2,$$

$$A_{ij} = \delta_{ij} \; , \quad B_{ij} = i \; \delta_{ij} + 1 \; ,$$

$$\delta_{ij} = \begin{cases} 0 & \text{if} \quad i \neq j \\ 1 & \text{if} \quad i = j \; . \end{cases}$$

The first four problems have symmetric objective functions.

Table (A.1)

Processing times for the sequential codes (time in seconds)

Function Starting Point $\underline{x}^{(0)}$	Newton-Raphson CPU Time	Variable Metric			Conjugate Gradient		
		CPU Time	No. of Funct. Calls	No. of Grad. Calls	CPU Time	Iter-ation	No. of Funct. & Grad. Calls
Quadratic							
$(0,1,....)^T$	15.00	1.80	8	5	1.23	3	9
non-symmetric	16.60	2.10	12	7	1.18	4	9
Rosenbrock							
$(0,1,....)^T$	154.23	72.11	414	210	5.67	19	48
$(-1.2,1,....)^T$	145.97	103.35	683	292	8.36	27	72
non-symmetric	140.98	80.78	496	249	11.36	49	103
Powell							
$(3,-1,0,3....)^T$	112.47	53.16	298	142	5.91	22	49
non-symmetric	134.97	41.28	218	114	11.06	35	87
Trigonometric							
$(1/64,1/64,...)^T$	* F	66.18	74	43	37.56	14	27
non-symmetric	2604.31	286.65	455	191	78.68	24	52
Box(M)							
$(1,2,1,2,...)^T$	4181.86	1161.27	429	227	137.81	6	16
non-symmetric	7196.09	3194.39	1263	621	354.02	19	40

* overflow in function evaluation.

Table (A.2)

DAP processing times for '1-D' search [VERSION A]

Function	Starting point $\underline{x}^{(0)}$	initial α_1	h	$\in$	Iter.	DAP time (secs.)
Quadratic	$(0,1,....)^T$	1.0	0.1	0.0001	2	0.8919624
	non-symmetric	1.0	0.1	0.0001	2	0.803285
Rosenbrock	$(0,1,0,1,...)^T$	0.5	0.001	0.001	14	5.532199
	$(-1.2,1.0,...)^T$	0.5	0.001	0.001	20	8.258047
	non-symmetric	1.0	0.001	0.001	26	9.842461
Powell	$(3,-1,0,3,....)^T$	1.0	0.01	0.0001	5	1.746016
	non-symmetric	1.0	0.01	0.0001	3	1.248776
Trigonometric	$(1/64,....)^T$	1.0	0.001	0.001	12	9.647102
	non-symmetric	0.5	0.001	0.001	14	13.33931
Box (M)	$(1,2,1,2,....)^T$	0.5	0.01	0.0001	18	452.6223
	non-symmetric	0.125	0.01	0.0001	21	508.0827

Table (A.3)

DAP Processing times for '1-D' search [VERSION B]

Function	Starting point $\underline{x}^{(0)}$	initial α_1	h	$\in$	Iter.	DAP time (secs.)
Quadratic	$(0,1,....)^T$	0.5	0.1	0.001	2	0.79532
	non-symmetric	0.5	0.1	0.001	2	0.774928
Rosenbrock	$(0,1,0,1,....)^T$	0.5	0.001	0.001	8	3.60468
	$(-1.2,1.0,...)^T$	0.5	0.001	0.001	25	9.791973
	non-symmetric	0.125	0.001	0.001	34	14.340800
Powell	$(3,-1,0,3,....)^T$	1.0	0.01	0.001	16	5.338949
	non-symmetric	1.0	0.01	0.001	10	3.542872
Trigonometric	$(1/64,....)^T$	1.0	0.001	0.001	13	11.07125
	non-symmetric	0.5	0.001	0.001	14	13.379770
*Box (M)	$(1,2,1,2,....)^T$	-	0.001	0.001	-	F
	non-symmetric	-	0.001	0.001	-	F

* overflow in function evaluation

Table (A.4)

DAP processing times for '2-D' search [VERSION A]

Function	Starting point $\underline{x}^{(0)}$	initial α_1	h	$\in$	Iter.	DAP time (secs.)
Quadratic	$(0,1,.....)^T$	1.0	0.1	0.0001	2	0.985168
	non-symmetric	1.0	0.1	0.0001	2	0.966224
Rosenbrock	$(0,1,....)^T$	0.5	0.001	0.001	12	6.486191
	$(-1.2,1.0,...)^T$	0.5	0.001	0.001	11	5.86484
	non-symmetric	0.5	0.001	0.001	71	36.626720
Powell	$(3,-1,0,3,...)^T$	1.0	0.01	0.0001	4	1.929656
	non-symmetric	1.0	0.01	0.0001	3	1.500664
Trigonometric	$(1/64,....)^T$	0.5	0.001	0.001	11	11.10786
	non-symmetric	0.5	0.001	0.001	17	17.57672
Box (M)	$(1,2,1,2,...)^T$	0.25	0.01	0.001	5	130.9869
	non-symmetric	0.25	0.01	0.001	8	202.3719

Table (A.5)

DAP processing times for '2-D' search [VERSION B]

Function	Starting point $\underline{x}^{(0)}$	initial α_1	h	$\in$	Iter.	DAP time (secs.)
Quadratic	$(0,1,\ldots\ldots)^T$	1.0	0.1	0.001	2	0.968968
	non-symmetric	1.0	0.1	0.001	2	0.949536
Rosenbrock	$(0,1,\ldots\ldots)^T$	1.0	0.001	0.001	6	3.282704
	$(-1.2,1.0,\ldots)^T$	1.0	0.001	0.001	4	2.290592
	non-symmetric	0.5	0.001	0.001	25	12.886690
Powell	$(3,-1,0,3,\ldots)^T$	1.0	0.001	0.001	7	3.154320
	non-symmetric	1.0	0.001	0.001	7	3.164520
Trigonometric	$(1/64,\ldots\ldots)^T$	0.5	0.001	0.001	10	10.048290
	non-symmetric	0.5	0.001	0.001	15	15.523480
Box (M)	$(1,2,1,2,\ldots)^T$	0.25	0.01	0.001	4	107.820100
	non-symmetric	0.0625	0.01	0.001	10	251.929100

Table (A.6)

DAP processing times for '4-D' search

Function	Starting point $\underline{x}^{(0)}$	initial α_1	h	$\in$	Iter.	DAP time (secs)
Quadratic	$(0,1,\ldots.)^T$	1.0	0.1	0.0001	2	1.237488
	non-symmetric	1.0	0.1	0.0001	2	1.210280
Rosenbrock	$(0,1,\ldots.)^T$	1.0	0.001	0.001	6	4.279285
	$(-1.2,1.0,\ldots)^T$	1.0	0.001	0.001	7	4.949902
	non-symmetric	0.5	0.001	0.001	41	28.211770
Powell	$(3,-1,0,3,\ldots.)^T$	1.0	0.01	0.0001	4	2.52372
	non-symmetric	1.0	0.01	0.0001	3	1.920984
Trigonometric	$(1/64,\ldots.)^T$	1.0	0.001	0.001	12	13.998930
	non-symmetric	0.5	0.001	0.001	7	8.657191
Box (M)	$(1,2,\ldots.)^T$	0.5	0.01	0.001	4	107.9466
	non-symmetric	0.125	0.01	0.001	13	323.9529

Table (A.7)

Performance measurement [VERSION A]

Define 'speed-up' ratio, $= \frac{\text{processing time using a sequential system}}{\text{processing time using the DAP}}$

Function	Starting point $\underline{x}^{(0)}$	Newton-Raphson DAP			Variable-Metric DAP			Conjugate-Gradient DAP		
		1-D	2-D	4-D	1-D	2-D	4-D	1-D	2-D	4-D
Quadratic	$(0,1,....)^T$	16.8	15.2	12.1	2.0	1.8	1.4	1.4	1.2	0.99
	non-symmetric	20.7	17.2	13.7	2.6	2.2	1.7	1.5	1.2	0.97
Rosenbrock	$(0,1,...)^T$	27.9	23.8	36.0	13.0	11.1	16.8	1.0	0.9	1.3
	$(-1.2,1.0,...)^T$	17.7	24.9	29.5	12.5	17.6	20.9	1.0	1.4	1.7
	non-symmetric	14.3	3.6	5.0	8.2	2.2	2.9	1.1	0.3	0.4
Powell	$(3,-1,0,3,...)^T$	64.4	58.3	44.6	30.4	27.6	21.1	3.4	3.1	2.3
	non-symmetric	108.1	89.9	70.3	33.1	27.5	21.5	8.9	7.4	5.8
Trigonometric	$(1/64,....)^T$	-	-	-	6.9	5.9	4.7	3.9	3.4	2.7
	non-symmetric	195.2	148.2	300.8	21.5	16.3	33.1	5.9	4.5	9.1
Box (M)	$(1,2,....)^T$	9.8	31.9	38.7	2.7	8.9	10.8	0.3	1.1	1.3
	non-symmetric	14.2	35.6	22.2	6.3	15.9	9.9	0.7	1.7	1.1

Table (A.8)

Performance measurement [VERSION B]

Define 'speed-up' ratio, $= \frac{\text{processing time using a sequential system}}{\text{processing time using the DAP}}$

Function	Starting point $\underline{x}^{(0)}$	Newton-Raphson DAP		Variable Metric DAP		Conjugate gradient DAP	
		1-D	2-D	1-D	2-D	1-D	2-D
Quadratic	$(0,1,....)^T$	18.9	15.5	2.3	1.9	1.5	1.3
	non-symmetric	21.4	17.5	2.7	2.2	1.5	1.2
Rosenbrock	$(0,1,...)^T$	42.8	47.0	20.0	22.0	1.6	1.7
	$(-1.2,1.0,...)^T$	14.9	63.7	10.6	45.1	0.9	3.6
	non-symmetric	9.8	10.9	5.6	6.3	0.8	0.9
Powell	$(3,-1,0,3,...)^T$	21.1	35.7	10.0	16.9	1.1	1.9
	non-symmetric	38.1	42.7	11.7	13.0	3.1	3.5
Trigonometric	$(1/64,....)^T$	-	-	6.0	6.6	3.4	3.7
	non-symmetric	194.6	167.8	21.4	18.5	5.9	5.1
Box (M)	$(1,2,....)^T$	-	38.8	-	10.8	-	1.3
	non-symmetric	-	28.6	-	12.7	-	1.4

REFERENCE INDEX

Ahmed, H.M. 8, 9
Ames, W.G. 45
Arun, K.S. 10
Aziz, K. 118

Ball, A.G. 135
Ball, W.W.R. 218
Barbacci, M.R. 31
Berzins, M. 4, 13, 16, 20, 25, 26, 29
Bojanczyka, A. 16
Bokhari, S.H. 10
Bowgen, G.S.J. 227
Brent, R.P. 16
Brodlie, K.W. 210, 222
Buckley, T.F. 4, 13, 16, 20, 25, 26, 29
Bye, C. 140

Calahan, D.A. 45
Calvert, W. 12, 25, 29
Campbell, R.H. 26
Carling, J.C. 118
Catt, I. 39
Cherry, L.L. 139
Clarke, M.R.B. 209, 213
Cohen, D. 16
Conway, L. 5, 9, 14, 17, 35
Curtis, A.R. 93

Davis, A. 16
Delosme, J. 8, 9
Delves, L.M. 167, 168, 185, 186, 187, 191
Denyer, P.B. 8
De Ruyck, D.M. 5
Dew, P.M. 4, 13, 16, 20, 25, 26, 29
Dixon, L.C.W. 229
Duff, M.J.B. 36

Eastwood, J.W. 55
Edmundson, H.P. 136
Eisenstat, S.C. 3
Enderby, J.A. 94

Finney, J. 8
Flanders P.M. 35, 135, 147, 206
Fletcher, R. 230
Flynn, M. 138
Forsythe, G.E. 222, 223
Froelich, R. 93

Gal-Ezer, R.J. 10
Gannon, D.B. 16
Gentleman, W.M. 16, 216, 225
Golub, G.H. 209, 225
Gostick, R.W. 178

Hageman, L.A. 92
Hall, C.A. 167, 185, 186
Hammarling, S.A. 220
Hanson, R.J. 75
Haynes, L.S. 2, 3
Heller, D.E. 16
Hellier, R. 168, 178
Hellums, J.D. 118
Henderson, W.D. 116
Hendry, J.A. 167, 168, 185, 191
Henrici, P. 210, 222, 223
Hestenes, M.R. 209
Higbie, L. 73
Hockney, R.W. 35, 36, 45, 55, 56, 59, 60, 122
Holstein, H. 130
Hunt, D.J. 135, 147

Ibbett, R.N. 7
Ipsen, I.C.F. 16

Jesshope, C.R. 35, 36, 45, 56, 60
Johnson, D. 94
Johnsson, L. 16
Jordan, H.F. 3, 11, 12,14, 25, 29, 30

Kahan, W. 209
Kincaid, D.R. 75
Kolstad, R.B. 26
Knowles, J.A. 94
Krogh, F.T. 75
Kuck, D.J. 16, 53, 209, 211, 212
Kung, H.T. 2, 3, 6, 7, 8, 11, 16
Kung, S.Y. 10

Lau, R.L. 2, 3
Law, G.T. 108
Lawson, C.L. 75
Leiserson, C.E. 7
Love, H.H. 3
Luhn, H.P. 136
Luk, F.T. 209

Mackinson, G.J. 147, 157
Manning, F.B. 38
Mathis, B. 136
McCallien, C.W.J. 92, 93
McKeown, J.J. 178, 182
Mead, C. 5, 9, 14, 17, 35
Mizell, D.W. 2, 3
Modi, J.J. 209, 213, 219, 223, 226, 227
Mohamed, J.L. 167
Moore, W.R. 36
Morf, M. 8, 9
Morjaria, M. 147, 157

Niblett, B. 140

Oldfield, D.E. 139
O'Neill, T. 167

Paddon, D.J. 130
Parkinson, D. 128, 135, 147, 157, 209
Pawley, S. 1
Peterka, V. 8
Petersen, W.P. 75
Phillips, C. 167, 185, 187
Podsiadlo, D.A. 3, 11, 14, 25, 30
Potter, D. 55
Powell, M.J.D. 210, 222
Price, N.H. 140
Pryce, J.D. 209, 219

Reddaway, S.F. 122, 135, 147
Reinsch, C. 225
Richardson, T. 94
Ridgeway-Scott, L. 4
Robinson, I.N. 36
Rogers, M.H. 9
Rush, J.E. 136
Ruitishauser, J. 220

Samba, A.S. 178, 191
Sameh, A.H. 16, 209, 211, 212
Sawyer, P.L. 11, 30
Saxe, J.B. 7
Scalabrin, M. 12, 25, 29
Schultz, M.H. 3
Seitz, C.L. 2
Sherriffs, V.S.W. 93
Siewiorek, D.P. 2, 3
Smith, B.T. 41
Smith, G.D. 121
Snyder, L. 4, 5, 25
Spalding, D.B. 118
Stansfield, E. 107, 108

Swarztrauber, P.N. 56

Sydow, P.J. 73

Temperton, C. 59

Thomas, G. 1

Trier, W. 116

Unruh, J.D. 5

Vestman, W. 139

Wachpress, E.L. 92, 93

Walkden, F. 108

Webb, S.J. 121

Weiser, U. 16

Whiteway, J. 178, 191

Wilkinson, J.H. 222

Wilkinson, J.M. 39

Wilson, A. 125

Wilson, E. 135

Young, C.E. 136

INDEX

accumulator, 36

activity control, 36, 38

alternating direction implicit (ADI), 121, 123, 124

ANALOGIC AP400, 68, 69, 71

APAL, 139

apparent parallelism, 45

APPLIED DYNAMICS ADIO, 69, 71

architecture
- general purpose parallel, 3
- lattice, 4, 25, 26
- parallel, 2
- systolic array, 5, 7, 9

array processing
- FORTRAN, 41-44
- FORTRAN examples, 42, 43

assembly language, 68
- CRAY-1, 74

banded matrix, 6, 95, 96, 167-183

bit plane, 138

bit mask, 139

bottleneck, 5, 18, 26, 29

carry bit, 36, 38

CDC 7600, 115, 123, 131, 185, 190, 191, 193

chaining, 7, 8, 13, 68, 74, 75

chessboard ordering, 121-124

CHiP, 4, 25

CLIP, 35, 36

CMU systolic system, 7, 8, 13, 14

CM*, 3

communication, 2-7, 12, 14, 25, 26
- ICL DAP, 136, 157

computation wavefront, 10

computational fluid mechanics, 107

computer performance, 46, 47
- apparent parallelism, 45
- pipelined computers, 47
- processor arrays, 48
- serial computers, 47
- two-parameter description, 46, 50

concurrent programming language, 26

conjugate gradient method, 11, 12, 14

conjugate gradient algorithm, 230, 237, 239

CRAY-1, 1, 7, 45, 50, 51, 54, 58, 59, 60, 62, 63, 69, 71, 73-89, 91, 93, 121, 123, 124, 131
- assembly language, 74
- double precision, 75
- FORTRAN, 73-89
- FORTRAN compiler, 74
- hardware features, 74
- instruction issue, 74
- memory conflicts, 73
- software considerations, 75

Crout factorization, 80

CSPI MAP, 69, 72

CYBER 205, 50, 51, 58-63

cyclic ordering, 209, 222, 224

cyclic reduction, 168
- block, 178
- block parallel, 56, 168, 171
- parallel, 168-178, 191
- point, 178

data crinkling, 112, 113, 128

data flow, 3

delay queue, 17, 18

digital control, 8

direct matrix method, 13, 14
- Gaussian elimination, 13, 14, 100, 101, 122, 192
- Gauss-Jordan elimination, 100, 168, 170, 171, 178, 190

document abstracting, 135-146

double precision, 75

elliptic partial differential equations, 167, 185

heat transfer, 118
 continuity equation, 118
 energy equation, 119
 momentum equation, 118
 vorticity equation, 118

FACR(ℓ) algorithm, 45

Fast Fourier transform (FFT), 57, 58, 60, 74, 122, 167, 187, 195-207
 discrete, 195, 196, 201
 parallel algorithm, 198
 8-point, 195, 199, 201, 205
 64-point, 200, 205, 206

fault tolerance, 39

finite element machine, 1, 11, 12, 13, 25, 29, 30

finite element method, 16, 91, 94, 167, 185

finite difference method, 18, 91, 108, 110, 119

finite volume, 108, 110

fixed point arithmetic, 8, 155

floating point computation, 8

floating point: 64-bit processors, 35

FORTRAN, 41-44, 69
 array processing example, 41-44
 CRAY-1, 73-89

FORTRAN: ICL DAP, 115, 120, 132, 157, 160, 195, 201-206, 210, 237
 curl operator, 127
 differential operators, 125
 div operator, 127
 grad operator, 126
 Laplacian operator, 127

FORTRAN compiler, 69, 70

FPS AP120B, 67, 70

FPS 164, 70

full adder, 37

Gaussian elimination, 13, 14, 100, 101, 122, 192

Gauss-Jordan elimination, 100, 168, 170, 171, 178, 190

Gauss-Seidel iteration, 13

GEC GRID, 36

Given's method, 16

Given's ordering, 209

Given's rotation, 224

Given's transformation (square root free), 216

glass-melting tank furnaces, 115-133

global bus, 14, 27, 29, 32

global element method (GEM), 167-183, 185-194

global matrix, 97, 98, 103-105, 185

Golub-Reinsch algorithm, 209

GOODYEAR MPP, 35, 36

hardware numerical library, 7

hessian matrix, 230, 232, 235

Householder's method, 16

Householder's transformation, 209

ICL 2980 series, 91, 93, 209

ICL DAP, 1, 4, 35, 36, 45, 50, 51, 54, 58, 59, 70, 71, 99, 107-114, 115-133, 135-146, 147-156, 157-166, 167-183, 185-194, 195-207, 209-228, 229-249
 bit masks, 139
 bit planes, 138
 data crinkling, 112, 113, 128
 floating point operations, 155
 FORTRAN, 115, 120, 132, 157, 160, 195, 201-206, 210, 237
 logical mask, 125, 132, 159, 181, 195, 198, 199, 205, 214
 logical vector, 138
 storage planes, 148, 150, 151, 154, 161, 163, 164, 166
 text storage, 138
 tridiagonal matrix operations, 124
 vector and differential operators, 125-127

ICL PERQ, 3, 4

ILLIAC-IV, 209, 224

INMOS TRANSPUTER, 2

inner product, 5, 12, 13, 14, 18, 25, 31, 79

inner product hardware, 12

Jacobi iteration, 1, 13, 16

Jacobi method, 219, 223

Jacobi parallel method, 209

Kalman filter, 8

Laplace's equation, 18

least squares method, 8

lexicon, 136

local data, 27, 29

local memory, 4

logical mask, 125, 132, 159, 181, 195, 198, 199, 205, 214

logical vector, 138

linear algebra (NAG BLAS), 75

LSI, 1, 10

LSOR, 123, 124

LU decomposition, 10

matrix, banded, 6, 95, 96, 167-183

matrix computation, 5, 9, 10

matrix element annihilation, 211, 219, 220
  scheme of Sameh and Kuck, 212, 213

matrix multiplication, 166, 178-182

matrix vector multiplication, 147-155, 157-166
  Parkinson's method, 154

memory conflicts, 73

Monte Carlo method, 93

multiplexing, 37

multiprocessor lattice, 10, 13, 14, 16, 21, 25, 26, 28, 29

multiprocessor lattice architecture, 4, 25, 26

NAG, 73-89
  basic linear algebra subprograms, 75

network simulation, 13

Newton's method
  grid search, 231, 234, 239
  modified algorithm, 229, 230
  modified Newton Raphson routine, 237, 239
  parallel algorithm, 230, 239

NMOS, 38

nuclear reactor, 91

paracomputer, 46

parallelism, measurement of apparent parallism, 45

Pascal compiler, 69

path Pascal, 25-33

path Pascal processes, 27

Parkinson's method, 154

partial differential equations, 18, 107

pipelining, 7, 10, 16, 17, 74

Poisson's equation, 45, 55

Poisson solution by:
  direct methods, 55
  blockcyclic reduction, 56
  FACR(ℓ) algorithm, 56
  ICL DAP FORTRAN, 127
  SERIFACR algorithm, 56
  PARAFACR algorithm, 60
  SERIFACR/PARAFACR comparison, 63, 64

precision:
  16-bit, 20
  double, 75

PRIME 750, 123, 124

processor array, 35-40

processing cells, 5, 20

processing element, 4, 10, 11, 12, 14, 30, 32, 35, 36, 37, 136, 138, 159

product planes, 160, 161

program development for array processors, 67

programmable switches, 4

programmable systolic chip, 6

QR algorithm, 16, 209

QU factorization, 210, 226
  parallel method, 212, 225

reactor design, 91-106

reconfigurable processor array, 35-40

reconfiguration control, 36-37

robotics, 7

Rosenbrock function, 236, 240

Sameh and Kuck annihilation scheme, 212, 213

scaler product, 11

self tuning control algorithm, 8

self tuning controllers, 9

signal processing, 8

single bit processor, 35

single bit processor array, 35

single value decomposition, 209-228
   algorithm, 217
   parallel method, 218, 224
   parallel algorithm, 218

slow down, 53, 54

software for array processors, 67-72

SOR, 93, 121, 122, 124, 128

sparse banded matrix, 16, 18, 147, 154, 155, 158, 166

sparse equation solver, 99

sparse matrix
   product, 166
   structured, 147, 155
   unstructured, 147, 154, 155, 157-166

special purpose processor, 3, 7, 9, 14

speed up, 53, 54, 80

storage planes, 148, 150, 151, 154, 161, 163, 164, 166

stress analysis, 95

syntax analysis, 135, 136, 139

systolic array, 1, 5, 6, 7, 8, 18
   processor, 5, 7, 8, 13, 16
   architecture, 5, 7, 9
   CMU system, 7, 8, 13, 14
   processor chip, 2
   programable chip, 6

square root filter, 7

text handling, 135

text parsing, 139

time complexity, 10, 18

tridiagonal matrix operations, 132

unconstrained optimisation, 229
   test problems, 240-249

variable metric algorithm, 230, 236

vector operators, 128

VLSI, 1, 5, 9, 25, 35, 36

VLSI parallel architecture, 2

wafer scale design, 38

wavefront, 10

wavefront array processor, 10

word parallelism, 35